# HIDDEN DIMENSIONS

COLUMBIA SERIES IN SCIENCE AND RELIGION

## THE COLUMBIA SERIES IN SCIENCE AND RELIGION

The Columbia Series in Science and Religion is sponsored by the Center for the Study of Science and Religion (CSSR) at Columbia University. It is a forum for the examination of issues that lie at the boundary of these two complementary ways of comprehending the world and our place in it. By examining the intersections between one or more of the sciences and one or more religions, the CSSR hopes to stimulate dialogue and encourage understanding.

Robert Pollack
*The Faith of Biology and the Biology of Faith*

B. Alan Wallace, ed.
*Buddhism and Science: Breaking New Ground*

Lisa Sideris
*Environmental Ethics, Ecological Theory, and Natural Selection:
   Suffering and Responsibility*

Wayne Proudfoot, ed.
*William James and a Science of Religions: Reexperiencing The Varieties
   of Religious Experience*

Mortimer Ostow
*Spirit, Mind, and Brain: A Psychoanalytic Examination of Spirituality
   and Religion*

B. Alan Wallace
*Contemplative Science: Where Buddhism and Neuroscience Converge*

Philip Clayton and Jim Schaal, editors
*Practicing Science, Living Faith: Interviews with Twelve Scientists*

B. Alan Wallace

# HIDDEN
# DIMENSIONS

**THE UNIFICATION**

**OF PHYSICS**

**AND CONSCIOUSNESS**

Columbia
University
Press ■
New York

Columbia University Press

*Publishers Since 1893*

New York   Chichester, West Sussex

Copyright © 2007 Columbia University Press

All rights reserved

Library of Congress Cataloging-in-Publication Data

Wallace, B. Alan.

Hidden dimensions : the unification of physics and consciousness /
   B. Alan Wallace.

   p. cm. — (The Columbia series in science and religion)

Includes bibliographical references and index.

ISBN 978-0-231-14150-5 (cloth)—ISBN 978-0-231-14151-2
   (pbk.)—ISBN 978-0-231-51220-6 (e-book)

1. Consciousness. 2. Quantum theory. 3. Physics—Philosophy. 4.
   Psychophysics. 5. Buddhism. I. Title.

BF311.W26668   2007

126—dc22

2006036326

# CONTENTS

# PREFACE AND ACKNOWLEDGMENTS

OVER THE past 400 years, the physical sciences have undergone two great revolutions, the first beginning with Copernicus and the second beginning at the turn of the twentieth century with the development of the theories of quantum mechanics and relativity. Since the mid-nineteenth century, the life sciences have been radically transformed by one great revolution, initiated by Charles Darwin. In contrast to those two fields of objective science, the mind sciences, which first appeared in the late nineteenth century, have yet to produce a single great revolution. One could say that the Copernican revolution took roughly 150 years to come to completion in the laws of classical physics formulated by Isaac Newton, and the Darwinian revolution took about the same time to come to fruition in the Human Genome Project at the beginning of the twenty-first century.

The second revolution in physics, however, has not been completed, for no one has successfully unified the great insights of quantum physics and the general theory of relativity. Fundamental problems remain. One of the central unsolved mysteries is the measurement problem, which has to do with the nature and significance of making a measurement of a quantum system. Before such a measurement, or observation, occurs, a quantum system is described in terms of abstract wave functions, or probability waves. Particles, such as electrons and photons, have no definite location and in fact do not even exist as discrete entities unless and until they are measured—they exist only as mathematical abstractions. Yet somehow these nebulous entities are measured with instruments of technology, with which they causally interact. Then these intangible quantum phenomena turn into the objectively real, elementary building blocks of the physical universe. No one yet knows how this transition from mathematical abstraction to concrete reality takes place, but in some way the observer—the person who designs and conducts experiments—plays a key role in bringing the quantum world to life.

Things get even stranger when quantum mechanics, a theory of the subatomic realm, is applied to cosmology. According to the equations of the new field of quantum cosmology, without reference to an observer, the universe as a whole is frozen into immobility. Physicists try to solve this so-called time problem by dividing the world into two domains: a subjective observer with his clock and other measuring devices and the rest of the objective universe. But it turns out that the quantum mechanical wave function of the rest of the universe depends on the designated time of the observer. And the notion of an observer necessarily implies the presence of consciousness, without which no observation ever takes place.

So quantum mechanics implies that consciousness may play a crucial role in the formation and evolution of the universe as we know it. But most researchers in psychology and brain science regard consciousness as nothing more than an emergent property of the brain, with no significance for the universe at large. The fundamental assumptions about the nature of the mind according to modern science are largely rooted in the mechanistic worldview of classical physics that dominated the late nineteenth century. And even today, students of the cognitive sciences are generally not required to study twentieth-century physics. The widespread, virtually unchallenged assumption in the discipline is that neither quantum mechanics nor relativity theory is relevant to the macroscopic, slow-moving phenomena in the brain that are relevant to the mind.

Many scientific studies indicate that mental phenomena—such as subjectively experienced desires, thoughts, emotions, and memories—influence brain function and behavior. In response to this empirical evidence, a growing number of cognitive scientists conclude that mental phenomena are real, but they insist that in order to causally interact with the brain, the mind must be physical. However, subjectively experienced mental phenomena lack any physical characteristics and cannot be detected with any of the physical instruments of technology, even though many specific brain functions have been identified that causally contribute to the generation of mental processes. Some scientists and philosophers of mind envision brain functions as having a dual identity, as both objective physical processes and subjective mental events. But they offer no explanation of what about the brain enables it to generate or even influence mental events, let alone allows specific neural processes to take on this dual identity. This is the so-called hard problem, and it has been unresolved since scientists first began studying the mind. Mental phenomena remain as much an enigma to cognitive scientists as the observer is to modern physicists.

A central hypothesis of this book is that the measurement problem in quantum mechanics, the time problem in quantum cosmology, and the hard problem in brain science are all profoundly related. If this is true, it implies that a solution to any one of them requires a solution to the other two. Chapter 1 sets forth the proposition that the mind sciences have failed to mature to the point of a revolution because they have failed to adopt a fundamental strategy that has been key to the success of physics and biology. While physicists and biologists have devised highly sophisticated means of directly observing physical processes and living organisms, cognitive scientists have failed to develop rigorous ways of directly observing mental phenomena. This exclusion, or at least marginalization, of subjectively experienced mental events from objective observation has resulted in a "blind spot" in the scientific view of reality.

Scientists' insistence that consciousness and all other mental phenomena must be physical is rooted in a naturalistic metaphysical framework, which maintains that only physical processes exert causal influences in nature. In chapter 2, various interpretations of naturalism are examined, leading to the startling conclusion that no one really seems to know what is meant by "physical"! While neuroscientists commonly regard this as an unproblematic issue, the more deeply physicists probe the nature of mass-energy and space-time, the more elusive the concept of matter becomes. Particularly in quantum physics, the objective, physical status of the material world independent of any system of measurement appears highly suspect.

Chapter 3 develops a more natural theory of human consciousness based not on the outdated assumptions of classical physics but in response to some of the keenest insights of contemporary physicists, including Freeman Dyson, John Wheeler, Paul C.W. Davies, Andrei Linde, and Michael B. Mensky. A central premise of this theory is that quantum physics, despite mainstream assumptions to the contrary, has great relevance to understanding mind-brain interactions and the role of the mind in the universe.

While astronomers have developed and refined the telescope to explore the depths of space and biologists have used microscopes to probe the nature of cells and genes, sophisticated means of exploring the space of the mind and the whole range of mental phenomena have yet to play a role in science. Chapter 4 presents methods for developing just such a "telescope for the mind," beginning with the meditative refinement of attention and introspection. Problems and solutions regarding the possibility of including introspection as an integral feature of the scientific study of the mind are then discussed.

Chapter 5 presents a "special theory of ontological relativity," proposing that mental phenomena do not emerge from the brain, but rather all mental and physical processes arise from another dimension of reality that exists prior to the bifurcation of mind and matter. Early versions of this hypothesis are traced back to Pythagoras and Plato, followed by a discussion of such a theory formulated by physicist Wolfgang Pauli and his colleague Carl Jung. Other, more recent physicists' related hypotheses, including those of David Bohm, Eugene Wigner, Bernard d'Espagnat, Leonard Susskind, Roger Penrose, and George Ellis, are also discussed.

As intriguing as these theories are, none of the above philosophers and scientists has been able to present any empirical means to put his hypotheses to the test. Chapter 6 takes the unprecedented step of proposing an array of experiments in consciousness that could be used to test scientific hypotheses of an archetypal realm of pure ideas. These experiments are based on ways of training the mind and experientially exploring the "form realm," in accordance with the meditative tradition of early Theravāda Buddhism of Southeast Asia. This chapter concludes with a discussion of the potential interface between such contemplative science and modern science as it has developed in the West.

Chapter 7 extends the theory of relativity already discussed to an all-inclusive, relativistic hypothesis about the participatory nature of reality, beginning with a discussion of related ideas by modern philosophers such as Ludwig Wittgenstein, Willard Quine, Hilary Putnam, and Bas van Fraassen and moving to provocative hypotheses of leading physicists, including Stephen Hawking, Gerard 't Hooft, John Wheeler, Anton Zeilinger, Hugh Everett, and Michael Mensky. A recurrent theme is the notion of the participatory universe as a self-excited circuit. These ideas are then compared to the Buddhist theory of ontological relativity known as the Middle Way philosophy, which is traced back to Indian Mahāyāna Buddhism in the second century.

As interesting as these philosophical and scientific theories are, physicists acknowledge that they have not been able to put them to the test of experience. Here again, the meditative tradition of Buddhism offers practical ways to explore the world of ontological relativity through highly advanced contemplative practices. These are explained in chapter 8, followed by a scientific evaluation of the credibility of such means of inquiry.

The final chapter of this book focuses on the theme of symmetry, which is central to modern physics. In particular, we return to the field of quantum cosmology and the problem of frozen time, in which the role of the observer again appears to be fundamental to the evolving universe. Beginning with a scientific discussion of this theory, we move to a meditative tradition that many regard as the pinnacle of Buddhist theory and practice, known as the Great Perfection, which is emphasized in the Vajrayāna Buddhism of Tibet. Examining the parallels between the scientific concept of the "melted vacuum" and the Buddhist theory of the absolute space of phenomena, this chapter sets forth the theory and practice of the Great Perfection and concludes with a discussion of complementarity between science and religion at large.

I would like to thank Arthur Zajonc, my principal mentor in physics, as well as Victor Mansfield and Michael B. Mensky for their helpful comments on this manuscript. I am deeply indebted to Wendy Lochner, the religion and philosophy editor at Columbia University Press, for her unflagging support of my work, and I am especially grateful to Leslie Kriesel, Senior Manuscript Editor, for her excellent work in editing this manuscript. I would also like to thank Nancy Lynn Kleban for her superb job of proofreading the entire manuscript. And finally, as always, I would like to express my heartfelt gratitude to all my teachers, East and West, and to my family, for their guidance, love, and wisdom, which have enriched my life in more ways than I can express.

# HIDDEN DIMENSIONS

# 1

# THE UNNATURAL HISTORY OF SCIENCE

## Unnatural Origins

In the four centuries since the scientific revolution, scientists have empirically investigated the objective physical world. Philosophers have primarily resorted to reason, backed by empirical scientific research, in their quest to understand the subjective mental world and its relation to the objective world. And theologians have based their understanding of the transcendent world of divine revelation—including angels, heaven and hell, and the nature of the Trinity—on their faith in God and belief in the veracity of his word as revealed through the Bible.

During those formative centuries of modernity, scientists continually developed effective means of observing physical phenomena, crucial for their extraordinary progress in increasing consensual knowledge of matter, energy, space, and time. Philosophers achieved no comparable success in developing effective means of observing mental phenomena, and this is one reason they have failed to develop any comparable body of consensual knowledge. Nor have theologians devised empirical means to test the articles of their religious faith, and the credibility of religious beliefs has steadily eroded under the onslaught of scientific discoveries.

By the closing decades of the nineteenth century, a growing number of scientists and other intellectuals were coming to the conclusion that only physical phenomena—those successfully observed and understood by science—were real. It was at this point in history that the scientific study of the mind began, a full 300 years after the scientific revolution. Since philosophers and theologians had failed to fathom the nature of the human psyche and spirit, scientists were ready to step in and complete their understanding of the natural world by including the subjective mind that had produced all objective scientific knowledge.

The history of science is marked by competing perspectives on which individuals and traditions of the past are authorities regarding the nature of reality and the distinction between appearances and reality. These two issues have always been closely interrelated. During the late medieval period in Europe, the Bible was widely regarded (under pain of death) as an infallible authority on the whole of reality, Aristotle as infallible on the world of nature, and Euclid as infallible on the axioms and theorems of geometry. Despite the many incompatibilities between the Christian and the scientific worldviews, in the thirteenth century, Thomas Aquinas ingeniously synthesized them into a single, coherent perspective that dominated European thought until the Renaissance.

With regard to celestial phenomena—the sun, moon, planets, and stars—the mainstream intelligentsia of the scholastic era, from the thirteenth century to the sixteenth century, were solidly behind Ptolemaic astronomy, which was based on such Aristotelian principles as the perfect immutability of these objects and their movement in perfect circles. Appearances that corresponded to those principles, such as the apparent movement of the sun around the earth, were accepted at face value, whereas incompatible appearances, such as the occasional retrograde movement of planets, were regarded as misleading. Their true, or essential, movements had to be understood in terms of the perfectly circular motion explained by epicycles and eccentrics.

As more precise empirical observations were gradually made, more and more epicycles and eccentrics had to be conjured up to account for discrepancies between appearances and the Aristotelian principles of nature. Then Copernicus, without making any significant empirical discoveries of his own, suggested a different perspective on the appearances of the relative movements of the sun, earth, and planets. He proposed that the appearance of the sun moving around the earth was an illusion and devised a mathematical theory for a heliocentric configuration of celestial phenomena. His theory accounted for observed phenomena at least as well as the Ptolemaic theory, while shifting the distinction between appearances and reality. But Copernicus was a devout Christian living in an era when his own church was putting heretics to death and condemning them to eternal damnation. When faced with the choice of publish or perish, he opted to perish first and publish later, thereby avoiding scrutiny by the Inquisition and securing his blessed tenure in the hereafter.

Copernicus provided a plausible alternative to the Ptolemaic theory of celestial phenomena that accounted for the same appearances with greater mathematical economy and simplicity. But to many intellectuals of his time, this was insufficient reason for abandoning the safe scholastic fusion of biblical and Aristotelian authority. Prior to Copernicus, there was a striking discrepancy: *theorizing* about celestial phenomena was done by highly trained professionals—including mathematicians, philosophers, and theologians—while empirical *observations* of celestial phenomena were left largely to amateur sky gazers relying on their unaided faculty of visual perception. Even Tycho Brahe's meticulous observations, which provided Johannes Kepler with the empirical data he used to formulate his three laws of planetary motion, were based on naked-eye perception. But there seemed no need to refine the methods of observation, for appearances were thought to be largely misleading. Even if more precise methods were devised, the empirical data would still be illusory, just as the close interrogation of a clever, consistent liar would bring one no closer to the truth.

But not everyone in the sixteenth century was content with such absolute reliance on the received wisdom of past authorities. Tycho Brahe devised a number of ingenious methods for professionally observing the relative movements of the planets. The data he collected were meticulously analyzed by Kepler, who became persuaded of the truth of Copernicus's heliocentric theory and was forced to the conclusion that the planets moved in elliptical, not circular, orbits around the sun. The beauty and elegance of Aristotelian physics was challenged by empirical data, and the theoretical constructs of the Ptolemaic epicycles and eccentrics, which had won the absolute allegiance of generations of astronomers through the Middle Ages, were discarded as elegant fictions.

Kepler's findings remained controversial: even Galileo did not rally to his support. But in the scientific revolution that followed, Galileo's refinement of the telescope and its unprecedented use in precisely examining celestial phenomena were key. Some scholastic philosophers refused to corroborate his findings by gazing through the telescope, but the tide of history was against them. One by one, the long-held beliefs of Aristotle and a literal reading of the Bible were overthrown by researchers professionally trained to observe celestial and terrestrial phenomena. Although appearances in nature are still regarded in many ways as being illusory and misleading, progress in science has relied on the collaboration between professional observers and experimenters and professional theoreticians. This gave rise to the first scientific revolution, in the physical sciences, begun by Copernicus and completed by Isaac Newton.

Newton sought to formulate the physical laws of nature bearing the absolute certainty of Euclidean geometry, but those laws can be discovered only through precise and thorough qualitative and quantitative observations and measurements of physical phenomena. Mathematical theories alone do not define, predict, or explain the emergence of a physical universe. In the language of pure mathematics, such terms as "mass," "energy," "space," and "time" have no meaning. They acquire meaning only as they are used to describe observations of physical phenomena.

Charles Darwin's careful observations of biological organisms overthrew the literal reading of the Bible, which states that animal species were created by divine intervention within a relatively brief period of time in the recent past and have been fixed ever since. This second scientific revolution was an agonizing conclusion for the intelligentsia of his era, most of whom were devout Christians and Jews who had always relied on biblical authority. It may be said that this first and only revolution in the biological sciences is currently coming to an end with the completion of the Human Genome Project, which explains the mechanisms by which natural selection occurs.

While biologists seek to formulate biological laws of nature with all the credibility of physics, physical theories alone do not define, predict, or explain the emergence of living organisms in the universe. Moreover, biological laws of nature are discovered on the basis of precise and thorough qualitative and quantitative observations and measurements of living organisms, not through a quantitative examination of their physical constituents alone. In the language of physics, terms such as "life," "death," "health," and "illness" have no meaning. They acquire meaning only as they are used to describe observations of biological phenomena.

## The Unnatural Emergence of the Mind Sciences

We have yet to achieve even one revolution in the mind sciences comparable to those in the physical and biological sciences. In this regard, science is now facing its greatest challenge since Copernicus. Science can either devise novel methods for rigorously examining mental phenomena or continue to rely primarily on the study of the physical correlates of the mind, while mental phenomena themselves display none of the normal physical characteristics of matter, such as mass, velocity, impenetrability, and spatial extension and location.

It is a natural human tendency to regard only the phenomena we are attending to as real, and things we fail to notice as epiphenomenal or simply nonexistent. Scientists are no exception to this rule. Since science is based on quantitative, objective observation, mental phenomena, which are qualitative and subjective, have largely been overlooked or marginalized. Even when scientists have turned their attention to mental phenomena, they have largely done so by posing questions about their neural causes and behavioral effects. Hardly any progress has been made in observing such phenomena directly, in the only way possible: by means of first-person observation, or introspection.

Scientists acquire empirical evidence according to the kinds of questions they pose and the methods of inquiry they adopt. Until now, the questions and methods have been overwhelmingly objective and quantitative, which inevitably has produced an objective, quantitative view of the universe at large, including the mind. Likewise, since the early twentieth century, the questions and methods used to explore the mind have been almost universally embedded within a materialistic ideology that assumes that all mental phenomena are functions or emergent properties of the brain. This mode of theory-laden inquiry guarantees that the empirical data acquired will conform to the assumptions underlying the research.

The challenge facing modern science is to either discover the laws, or regularities, of mental phenomena in the same way it has explored physical and biological phenomena—by careful examination, with as few ideological assumptions as possible—or continue exploring the mind primarily by examining its physical correlates, which only reinforces the materialistic assumptions held during the late nineteenth century, when the scientific study of the mind began.

A true revolution in the mind sciences has been delayed by an enforced conformity to the unnatural ideological and methodological constraints imposed by the assumptions of scientific materialism, particularly neo-Darwinism. One such assumption is that mental phenomena are equivalent to neurophysiological processes in the brain, an empirically uncorroborated belief. If the first revolution in the mind sciences is to take place, such unsubstantiated ideas must be suspended and new methodologies must be employed that are uniquely suited to the scientific study of mental phenomena, including consciousness. In other words, science can either continue to let its study of the mind be dominated by the metaphysical assumptions of a well-established ideology or pursue the open-minded, empirical investigation of mental phenomena, even if it calls into question some of the most deeply held scientific beliefs based on classical physics and contemporary biology.

The major alternatives we have today as definitions of the nature of consciousness are that it is a supernatural phenomenon that operates according to laws outside of those governing the physical world or that it is a natural phenomenon, an attribute or emergent property of physical processes.

In the late nineteenth century, following three centuries of extraordinary success in the quantitative, objective study of physical phenomena, scientists took on the daunting challenge of studying mental phenomena through experimental psychology. After thirty years of ineffectively utilizing introspection in the scientific study of the mind, twentieth-century academic psychology (particularly in the United States) abandoned any attempt to develop rigorous means of observing mental phenomena. Researchers reverted to the time-tested, objective, quantitative methods of the physical and life sciences for studying the behavioral expressions of mental processes. Much can be learned by drawing inferences about *causal* mental processes on the basis of their *resultant* modes of behavior and verbal reports, as has been done in behaviorism and psychoanalysis. But radical behaviorists were driven to a more drastic claim stemming from their commitment to a materialist ideology: because mental phenomena, including consciousness, could not be physically measured, they should be deemed nonexistent! And anyone who insisted on affirming their own first-person experience of their thoughts, emotions, dreams, and perceptions was condemned for clinging to ancient superstitions and magic.[1]

## A Blind Spot in the Scientific Vision of Reality

Mental phenomena have always occupied a blind spot in the objective, quantitative scientific vision of reality, and since they could not be detected by the five physical senses or any of the measuring devices developed through advances in technology, behaviorists, equating scientific knowledge with human knowledge, simply denied what they could not observe in the laboratory. In an extraordinary triumph of ideology over experience, some insisted that all subjective terms, including "mind" and "ideas," be banned from scientific discourse. This categorical refusal to admit the existence of mental phenomena has filtered into mainstream academic philosophy, with some prominent thinkers denying the existence of subjective statements[2] and others maintaining that subjectively experienced mental states must be nonexistent, for the descriptions of such states are irreducible to the language of neuroscience.[3]

Nowadays most philosophers and cognitive scientists have distanced themselves from this extreme ideological commitment to materialism, which so obviously flies in the face of personal experience. Thoughts and mental images, desires and beliefs, emotions and dreams do exist, and somehow their awkward subjective presence must be incorporated into a scientific view of nature. All subjective experiences, including consciousness itself, remain invisible to objective scientific observation. A growing number of scientists and philosophers of mind believe they have the solution: simply declare that conscious states are equivalent to their neurophysiological correlates or to higher-level features of the brain.[4] In this view, conscious mental events occupy a unique status among physical phenomena. The physical processes in the brain that are equated with mental processes are believed to have a dual aspect: they are physically measurable processes, consisting of ordinary electrochemical events of a kind quite familiar to physicists and chemists, but somehow, inexplicably, they are also subjective experiences. The rationale for this quasi-dualistic position is that mental phenomena *appear* to be nonphysical, but this appearance is misleading, for they are *realized* as neural events, which are their *essential nature*.[5]

It is as if mental phenomena, despite their undeniably subjective, nonphysical appearance, are being granted admittance into the world of nature by being equated with well-understood physical phenomena. Scientists have yet to identify the neural correlates of consciousness, so no one even knows yet what those hypothetical neural processes with a dual identity might be. But advocates hold to this position for two reasons, one based on common sense and the other based on four centuries of scientific discoveries: in deference to common sense (which some behaviorists and eliminative materialists abandoned), they admit that mental phenomena do exist; and in light of the widespread scientific assumption that only physical phenomena exist and are causally effective in the natural world, they conclude that mental phenomena must be physical, *even if they don't appear to have any physical attributes and cannot be detected by any scientific instruments designed to measure all known types of physical phenomena*. To appreciate this point, one must recognize that the detection of the physical correlates of mental phenomena through brain-imaging and other kinds of technology is just that: measurement of physical *correlates* of mental phenomena, not of the mental phenomena *themselves*.

As science focuses its one good eye—the eye that detects objective physical phenomena—on nature, mental phenomena remain hidden in its blind spot. Scientists are doing what the brain does when presented with a blind spot corresponding to the point where the optic nerve touches the back of the retina: they cover the unknown contents with familiar phenomena that are proximate to the black hole. Physical processes closely correlated with mental phenomena are now called on to fill in, performing a double duty—subjective and objective—that is found nowhere else in the universe. Instead of discovering the nature of mental phenomena by carefully observing them, as has been done for all other kinds of natural phenomena, scientists are simply decreeing the equivalence of mental phenomena and their neural correlates, without any direct evidence.

While cognitive scientists seek to formulate the cognitive laws of nature with all the credibility of biology, biological theories alone do not define, predict, or explain the emergence of consciousness in the universe. In the language of biology, such terms as "desire," "attention," "emotion," and "consciousness" have no meaning except what they acquire on the basis of observations of mental phenomena. Likewise, psychology alone does not define, predict, or explain the emergence of philosophy, defined as the systematic, rational exploration of what we know, how we know it, and why it is important that we know it. Nor does philosophy alone account for the emergence of religious beliefs and experiences.

## Idols of Human Knowledge

Dualist and materialist theories of consciousness, for all their differences, have one trait in common: they do not lend themselves to empirical verification or repudiation. Those who assert that the mind is a nonphysical phenomenon that may exist independently of the brain have never been able to provide any supporting empirical evidence *by using the methods of mainstream scientific inquiry*. But neither have those who insist that the mind is either nonexistent or equivalent to brain functions. For all current means of scientific inquiry entail observations and experiments on physical processes, which precludes the very possibility of encountering any nonphysical mental events. As long as cognitive scientists continue to confine their observations to objective measurements of the neural causes and behavioral expressions of mental phenomena, neither the dualist nor the materialist hypothesis concerning the mind/body problem can be corroborated or repudiated. So neither dualist nor materialist theories of the mind are truly scientific. They are simply expressions of the ideological commitments of their adherents.

How are we to extricate ourselves from this morass of ideologically driven beliefs? Advocates of supernaturalism, including but not confined to religious fundamentalists, insist that scientific discoveries must be subservient to divine revelation. Advocates of scientific naturalism virtually define their view of the world by their refutation of the existence of supernatural entities, which include anything nonphysical. Each side of this controversy has great political and economic backing, and neither shows any indication of backing down.

To try to find a resolution to this stand-off, with each side hurling abuses when they are not actively seeking to annihilate the other, let us look back to the time when this controversy was first ignited. The late medieval era produced a relatively stable, scholastic integration of biblical theology and Aristotelian philosophy and science. Conformity to this worldview was enforced by the unified might of the Roman Catholic Church and the kingdoms ruling Europe. Medieval scholastics believed that human knowledge stemmed from two sources: the Bible, consisting of God's word, and nature, which was created by God. This implied an inevitable conformity between the two, with the former dictating how to read the latter. This gave rise to a strictly enforced hierarchy of knowledge:

theology

philosophy

science

A primary characteristic of this medieval hierarchy was a top-down insistence on conformity to an ideology considered to be essentially complete and perfect. All experiential findings, contemplative or scientific, had to conform to that ideology. It was against this imperative that the pioneers of the scientific revolution rebelled. One of the architects of this new, scientific view of nature was Francis Bacon, and he introduced a notion that is still pertinent. An idol, he declared, is the unaffected partner in the coupling of two phenomena.[6] There have been many occasions in the history of science when one natural phenomenon was thought to influence another without undergoing any reciprocal influence. To take a recent example, until Albert Einstein presented his general theory of relativity, scientists believed that matter affected space in the sense that a region of space could be filled with or emptied of particles, waves, and so on, but space did not exert any reciprocal influence on matter. Matter, as the unaffected partner in this coupling, took on the role of an idol. But Einstein's great insight was that matter curves space-time, and curved space influences how matter moves. This means that space does influence matter, which therefore is stripped of its status as an idol. Physicists today do not know of any phenomenon in which one subject is influenced by another without exerting an influence back.[7] Nature, it turns out, abhors idols.

The medieval hierarchy of knowledge was stacked with idols. Biblical theology, at the top of the totem pole, exerted an enormous influence on philosophy, including natural philosophy, but it was unthinkable for philosophers to rewrite or edit the Bible. In this coupling, theology became an idol. Likewise, Aristotelian philosophy dictated what kinds of scientific inquiry were viable, but, as Galileo found out the hard way, empirical discoveries that challenged either the Bible or Aristotle were forcibly repressed. So in the coupling between Aristotelian philosophy and empirical science, the former towered over the latter as an idol.

The medieval hierarchy of knowledge eventually collapsed from the bottom up due to generations of scientists making empirical discoveries, based on the close observation of natural phenomena, that clearly contradicted literal readings of the Bible and Aristotle. Since the scientific revolution, theologians have primarily emphasized belief as a means to understand the transcendent realities revealed in their scriptures. Philosophers have primarily relied on reason to unveil the secrets of the mind and its relation to the objective world of science. But scientists have let empirical evidence be the final arbiter of their theories. If a theory can't be tested empirically, theologians and philosophers may try to evaluate it, but it doesn't qualify as a scientific theory.

The history of science has shown that physical laws are discovered by observing physical phenomena, and biological laws are discovered by observing biological phenomena. It should follow that psychological laws are discovered by observing mental phenomena, and spiritual laws are discovered by observing spiritual phenomena. This was precisely the strategy proposed by William James when the scientific study of the mind and religion began.[8] But instead of following his lead, twentieth-century science adopted a new hierarchy of knowledge, replete with its own idols:

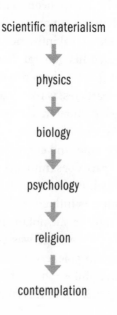

scientific materialism

physics

biology

psychology

religion

contemplation

A primary characteristic of this materialist hierarchy is a top-down insistence on conformity to a materialist ideology based on four fundamental assertions: the universe emerged solely as a result of physical events occurring at the time of the big bang, the mathematical principles of which are well understood by theoretical physicists; living organisms evolved solely from inorganic physical processes, the constituents of which are well understood by chemists; mental phenomena emerged solely from organic processes, the constituents of which are well understood by biologists; and religious beliefs and contemplative experiences emerged solely from mental processes, the constituents of which are well understood by psychologists. All empirical research in the natural sciences, with very few exceptions, is required to conform to those assertions.

In the above hierarchy, each of the higher fields of knowledge takes on the role of an idol for those beneath:

- Physicists do not challenge the principles of scientific materialism.
- Biologists do not challenge the principles of physics.
- Psychologists do not challenge the principles of biology.
- Scholars of religion do not challenge the principles of psychology.
- Contemplatives have no voice in academia, so it doesn't matter if they challenge anyone.

Despite the many successes of this physically reductionist hierarchy, this unilateral approach to knowledge leaves fundamental questions unanswered:

- Scientific materialism has no explanation for the mathematical nature of physical laws.
- Mathematical theories alone do not define, predict, or explain the emergence of a physical universe.
- Physical theories alone do not define, predict, or explain the emergence of life in the universe.
- Biological theories alone do not define, predict, or explain the emergence of consciousness in living organisms.
- Psychological theories alone do not define, predict, or explain the emergence of religious belief or contemplative experience in conscious beings.

Viewing the whole of reality through the one eye of objective scientific inquiry has left us in the dark especially regarding consciousness. After 400 years of breathtaking scientific progress, there is still no scientific definition of consciousness; no objective means of detecting consciousness; ignorance of neural correlates of consciousness; ignorance of the necessary and sufficient causes of consciousness; and ignorance of how the brain generates or even influences consciousness.

The natural evolution of the universe is now assumed to correspond to the unnatural evolution of human knowledge since the scientific revolution: physics → biology → psychology. The big bang is believed to have been a sudden spontaneous appearance of space-time from nothing, a transformation that took little more than a Planck time, or about $10^{-44}$ second. Life is likewise thought to be a sudden spontaneous appearance of biological organisms from inorganic matter. And consciousness is widely assumed to have emerged from highly interconnected networks of a large number of heterogeneous neurons.[9] But none of these three hypotheses has lent itself to empirical corroboration. There could have been other factors leading to the emergence of the physical universe. The emergence of life might have involved a sudden change or a long sequence of transitional states extended over millions of years. Nobody knows. And the emergence of consciousness in the universe and in an individual human being remains deeply problematic.[10] There is something unnatural about the way science has evolved, first operating under the ideological constraints of Roman Catholic theology and being maintained in its present incarnation by the constraints of scientific materialism.

## Naturalizing the Mind Sciences

For the first revolution in the mind sciences to take place, those who are existentially committed to the materialistic view of the mind will face an agonizing prospect. The Copernican theory relativized the position of the earth by displacing it from the absolute center of the universe to one of a cluster of planets. The Darwinian revolution likewise relativized man by displacing him from the status of a creature uniquely formed by God in his own image to a member of a family of primates produced by natural selection. Likewise, the first revolution in the cognitive sciences will relativize the human mind by displacing it from a physical function of the brain to an emergent process arising from a dimension of reality more fundamental than the duality of mind and matter. The implications for the rest of science are enormous.

Historically, science developed from physics to biology to psychology. In accordance with that sequence, biologists seek to understand living organisms in terms of physics and psychologists are intent on understanding the mind in terms of biology, with the latter domains of reality being epiphenomenal to the former. It is assumed, therefore, that discoveries in biology will have no impact on physics and those in psychology will have no impact on biology. But if the scientific mind is a part of nature and doesn't stand outside it, we should consider the fact that in the whole of nature, there is always a reciprocal effect in any coupling of two or more phenomena.[11] So if the scientific mind abides by the laws of nature, there should be reciprocal effects among all the sciences, unless they are obscured by ideological constraints.

The close correlation between the historical development of mathematics, physics, biology, and psychology and the widely held ontological hierarchy among these four fields of inquiry can hardly be coincidental. Did God guide the course of scientific inquiry so that it would parallel the fore-ordained, innate hierarchy of nature? In other words, has there been an intelligent design behind the evolution of the scientific worldview, or has science survived, adapted, and mutated over the centuries in response to a changing climate of human societies in dynamic interaction with their natural environment?

Taking the latter, naturalistic alternative as our working hypothesis, it would immediately follow that someone might have formulated Darwin's evolutionary theory before Copernicus, and someone else could have devised William James's introspection-based psychological theories before Darwin. To be sure, biology *as we know it* could not have evolved without a strong basis in physics, nor could psychology *as we know it* have evolved without a strong basis in biology. But the scientific disciplines of mathematics, physics, biology, and psychology could have evolved in different sequences, which would have resulted in different mutations of these modes of inquiry.

Traditionally, a scientific truth must fulfill two criteria: be free of subjective human biases and strongly accord with the natural world that exists outside our minds. Skepticism and empiricism have historically played the role of healthy predators in natural selection, enabling only the strongest theories to survive and procreate. In this evolutionary metaphor, mutations of novel theories and unprecedented modes of experiential inquiry play a crucial role, and researchers have attacked such new theories and observations to make sure they are free of subjective biases. But some species of maladaptive scientific thought may outlive their time, artificially preserved by institutions bent on maintaining ideological and methodological conformity. This occurred during the late medieval era under the auspices of the Roman Catholic Church, and it is occurring now under the influence of various institutions ideologically and economically committed to scientific materialism.

If natural selection and survival of the fittest is a good metaphor for the development of specific theories, it may also apply to the evolution of entire fields of scientific inquiry. But for a scientific theory to survive and procreate, it must adapt not only to the world of empirical research but also to the ideological and sociological world of the people conducting that research. Empirical evidence resulting from rigorous investigation may determine whether a particular theory survives scientific scrutiny, but it does not determine which kinds of questions scientists pose or the methods they use to answer them.

Scientists today are faced with the unique challenge of evaluating theories regarding mental phenomena, which exist within our minds, in relation to physical phenomena outside our minds. Hypotheses resulting from scientific inquiry must ultimately lead to testable consequences—even if it takes decades—if science is to advance. Otherwise, theorists are doing metaphysics, not science. All the current mainstream scientific theories regarding the nature of mental phenomena are based on the assumption that they are emergent properties or functions of matter. And all mainstream empirical research in this area accords with that assumption, so materialist theories of the mind are relieved of the requirement of leading to testable consequences.

The central theme of this book is that the multiple dimensions of the natural world, including consciousness and all objects of consciousness, can be understood only by focusing both a scientific vision and a contemplative vision on the world of human experience. There is no place for idols in this world, for nature abhors idols, including those stacked up in the unnatural hierarchies of both medieval and modern knowledge. No field of human knowledge is possible without consciousness, which is the foundation of our perceptual and conceptual knowledge of the universe. So in place of these outworn hierarchies, I propose the following dynamic lattice of knowledge. The scientific and contemplative study of consciousness is in the center, while reciprocally influencing all the fields around it; these, in turn, reciprocally influence one another.

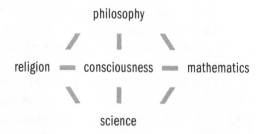

According to the current mainstream hierarchy, biology idolizes physics, the cognitive sciences idolize biology, the philosophy of mind idolizes the cognitive sciences, and the academic study of religion idolizes all the above. As a result of these asymmetrical relationships among domains of knowledge, we overlook the ways the latter may dynamically influence the former in each of the couplings.

The model proposed here entails a dynamic integration of the wisdom of the world's sciences, philosophies, and religions. The working hypotheses underlying this lattice structure are that no one religion, ideology, or civilization has a monopoly on truth, and that human knowledge of nature is continually evolving. To put this approach into action, we will have to throw down all idols and proceed with a spirit of true empiricism, questioning all ideologies, both religious and materialistic. By doing so, we have the opportunity to set in motion the first revolution in the mind sciences, and that may have deep implications for all other branches of science, in which the role of consciousness in nature has until now been ignored.

# 2

# THE MANY WORLDS OF NATURALISM

**MANY SCIENTISTS** and philosophers have recently turned their attention to understanding the nature of consciousness, and the great majority of them are determined to provide a "naturalistic" solution to the mind-body problem. But a variety of interpretations of naturalism have been advocated, so before evaluating the merits of specific views of consciousness, let us inspect the background range of perspectives on naturalism.

## The Natural World = The Physical World

According to one common interpretation, naturalism is a view of reality that excludes the possibility of nonphysical agents, forces, or causes. Understanding this requires a close look at the meaning of the terms "physical" and "matter." During the closing decades of the nineteenth century, when the scientific study of the mind began, a material body was defined as being permanently located in space, causally connected to changes in its spatial environment, and endowed with mass. But during the early decades of the twentieth century, advances in quantum mechanics challenged all three of those criteria, undermining the classical concept of matter as a collection of *inherently massive* and *spatially defined* particulate bodies.

Erwin Schrödinger, one of the pioneers of quantum mechanics, formulated a devastating critique of this classical notion,[1] and contemporary physicists and philosophers have updated the challenge to nineteenth-century materialism.[2] Nevertheless, many unsophisticated physicists still portray the world as made up of small (more or less localized) entities called "particles" that *contain* energy and *interact* with fields and waves, an antiquated picture left over from ancient atomism and nineteenth-century classical materialism. Some try to accommodate the classical notion of matter by describing particles as "wavicles" (a reified combination of a wave and a particle), but in this account, there is no vestige of a localized particle in space-time or of an even distribution of energy that can be ascribed to a wave. In other words, this attempt to salvage a classical material entity actually reduces matter to a conceptual abstraction. Indeed, many research physicists are finding that the more deeply they examine the nature of matter, the more elusive, mysterious, and mathematical it appears to be.[3]

Naturalism is also closely linked to the notion of energy, which took a firm place in the scientific worldview in 1847, when Hermann von Helmholtz treated mechanical phenomena, heat, light, electricity, and magnetism as different manifestations of energy, which he foresaw would serve as a unifying principle in physics. The actual term "energy" was first used as a general and fundamental concept by William Thompson two years later. He defined it as an objectively real, quantitatively immutable entity that is convertible and links all of nature in a web of energy transformations. Early explanations for the propagation of energy in the form of light, electricity, and magnetism required the existence of a luminiferous ether, a physical medium permeating space that would ripple when waves of energy passed through it. But in 1887, the existence of such a mechanical medium was disproved by the renowned Michelson-Morley experiment, which was meticulously designed to detect the luminiferous ether and came up with negative results. Since then, the very notion of electromagnetic fields that are propagated through space has been reduced to a set of mathematical abstractions. Although such fields exhibit wave properties, there is no material stuff in objective space that mechanically ripples, like waves in a pool of water. Further developments in twentieth-century physics have refined the general concept of the nature of energy to a mathematical principle, not a description of a mechanism or anything concrete. No one today really knows what energy *is*.[4]

## The Natural World = The Objective World

According to contemporary physics, all configurations of mass-energy are regarded as oscillations of abstract, mathematical fields in space, a far cry from the substance materialism of the ancient Greeks and classical physics![5] This has in no way diminished the naturalist commitment to the primacy of matter, which now operates as a symbolic commitment to *objective science*: matter represents objective reality existing outside the mind.[6] The fundamental premise in this interpretation of naturalism is that something is material if it may appear in space-time to anybody, and if its appearances are constrained by certain clauses of objectivity. This notion raises further problems from the perspectives of science and religion. First of all, in terms of scientific research, many sophisticated observations require years of rigorous training, from molecular biology to astronomy. Scientists need to learn what to look for before they can recognize what is presented to their physical senses, a principle that has been carefully researched by psychologists and neuroscientists for decades.[7] With training, and thorough familiarity with the theories of their areas of specialization, scientists learn to observe things in the objective world that are invisible, or at best unintelligible to the untrained observer. The "third person" in scientific research has never been a causal observer called in from the street, but a professional researcher with a set of assumptions shared by a given community.[8]

Second, in terms of religious experience, many people throughout history have allegedly observed a wide range of supernatural and immaterial phenomena that appeared to them in objective space. During the sixteenth century, there was a widespread acknowledgment that reports of witchcraft were based on the independent and concordant testimony of many witnesses.[9] And one of the most severe critics of magic, who claimed that the witches' Sabbath and flights to these nocturnal gatherings were diabolical illusions, found his own position weakened by the admission that he himself, in broad daylight, before an audience, had witnessed the levitation of a witch into the air.[10] Since then, there have been countless instances of people throughout the world claiming to have observed angels, ghosts, and a myriad of other supernatural phenomena. Some of these observations, like scientific observations, were restricted to a privileged few, while others were witnessed by the general public. Very few mainstream scientists today have deigned to subject such reports to rigorous scientific scrutiny, for the very possibility of observations of immaterial entities by anyone but a well-trained scientist is viewed with extreme skepticism. Scientists, using the latest instruments of measurement, may make unprecedented observations of novel phenomena, but such investigation is closed to the general public, and especially those with religious commitments.

There is yet a third problem with the proposal that something is material if it may be objectively observed or measured. Contemporary versions of superstring theory, or M-theory, posit seven hidden dimensions underlying the familiar three of space and a fourth dimension for time, in addition to an array of "superparticles" for which no empirical evidence has yet been discovered.[11] Are we to accept the material, or objective, existence of these entities without empirical evidence? On this question, leading physicists take opposite sides, some highly optimistic that M-theory will eventually be empirically validated[12] and others extremely skeptical.[13]

Whatever the future of M-theory, the twentieth-century revolution in relativity theory and quantum theory has already cast the notion of objectivity in a whole new light. Since the advent of the special theory of relativity in 1905, time and space can no longer be regarded as absolutely objective phenomena. It now makes no sense to imagine or search for an objectively true situation at a particular moment in time, existing "behind" relative appearances. The notions of before and after have no absolute or universal meaning; they take on particular meanings only to well-defined observers.[14] Moreover, electromagnetic fields and mass, previously thought to have absolutely objective existence, can be understood only within the framework of an observer-dependent theory of space-time. According to relativity theory, the things that are observer independent are not invariants that exist in the phenomenal space of our lived experience, but rather abstract mathematical spaces in four-dimensional space-time. But physicists know that even such invariants are not absolute according to the transformations specified by the general theory of relativity. So we cannot consider them to be ultimately objective features of reality either.

The ideal of objective knowledge of objective physical realities was further undermined by advances in quantum theory during the opening decades of the twentieth century. The wave function describing a quantum system does not exist in objective space and time, only in an abstract mathematical configuration. And in order to make predictions, an observer-participant must define his particular frame of reference and how he wishes to represent the wave function within it. Knowledge only arises in relationship among the observer, the system of measurement, and the measured object, so it no longer makes sense to think of attributes of an object apart from the actual experimental setup designed to measure them. The problem of objectivity is especially evident in cases that Erwin Schrödinger called "entanglement," in which the experimenter's frame of reference and the system under study can only be regarded as an inseparable whole. Although quantum physicists have formulated a theory, called "decoherence," that describes the set of alternative results of measurement and the associated probability distribution, they have yet to explain the mechanism of selection of one particular result. And such an explanation must somehow include the mind of the observer.[15]

## The Natural World = The World of Physics

One way of circumventing the above problems regarding the definition of matter is to declare that something is material that occupies or takes place in space, and consists of properties and relations, actions and interactions of particles and fields, *or whatever (as yet undefined) basic entities physics deals with.*[16] The responsibility of determining what is material is shifted away from notions of matter and objectivity to a particular community of scientists: naturalism includes in the category of "physical" whatever physicists say belongs there. Instead of appealing to standards of empirical observation or reason, this criterion simply grants the authority to determine what is and is not "natural" to a privileged group of people deemed to be ultimate authorities on the subject. It is just as legitimate to rely on physicists to determine what is physical as it is to rely on biologists to establish what is living and on psychologists to determine the scope of psychological phenomena. But the theories and methods of physics are far too limited to set the standard for what is "natural," especially when "natural entities" alone are deemed to be "real." The entire range of mental phenomena, for instance, is theoretically inexplicable and empirically undetectable as far as physics is concerned, but we do not turn a blind eye to them and deny their very existence.

A more nuanced definition along these lines is that naturalism includes only those kinds of causal interactions that can be seen, discovered, or inferred by way of known and reliable epistemic methods. This implies that scientists must use only the best, or most widely accepted, scientific epistemology and ontology to regulate their inquiry. Then naturalism is not so much about detecting a certain class of objective phenomena but about accepting a kind of "epistemic policy" that determines what counts as fact. This involves a stance, a way of behaving, an interpretive orientation, and a commitment to act and understand things according to a certain outlook.[17] There is a wide range of known and reliable epistemic methods for observing and analyzing material and biological phenomena, but what are the "acceptable epistemic methods" for observing mental phenomena? Generally speaking, the best scientific epistemology should be determined on the basis of the distinctive characteristics of the phenomena themselves, not on the basis of prior ideological commitments.

When seeking to understand a class of natural phenomena that is undetectable by all known scientific means, new epistemic methods must be explored and made as reliable as possible. For mental phenomena, introspection is the obvious candidate for inclusion as a scientifically viable epistemic method, because it is the only means we currently have of observing subjective events. If there are laws and regularities within the domain of mental phenomena, the history of science suggests that they will be discovered only through meticulous observation of those phenomena, just as physical laws were discovered through rigorous observation of physical phenomena. But scientific resistance to this approach remains strong, fueled in part by an ideological commitment to the principles of neo-Darwinism.

## The Natural World = The World of Neo-Darwinism

A final approach to naturalism consists of a commitment to the neo-Darwinian theory of evolution, according to which humans are mammals fully subject to the laws of physics and biology. An immediate implication of this stance is the conviction that it is only a matter of time before a comprehensive explanation of mental phenomena will be provided using the tools of neurobiology: by using well-known and reliable epistemic methods of studying the brain, biologists will eventually be able to explain satisfactorily all the steps that intervene from neural patterns to subjectively experienced mental events.[18]

Although scientists have no widely accepted theory about how certain organisms first became conscious or how and when a human fetus becomes conscious, most cognitive scientists are thoroughly convinced that given the success of Darwinian theory in explaining natural selection, it is the most promising for eventually explaining the origins and nature of consciousness in the universe. But it is a categorical error to assume that a scientific theory that explains one category of natural phenomena is equally capable of explaining other categories. An athlete may be a superstar in basketball, but that is no guarantee that he will excel in another sport, such as baseball.

Whatever the future successes or failures of neurobiology in explaining the nature and origins of mental phenomena, the fact remains that these phenomena are invisible to all objective means of scientific measurement. Consequently, one of the major neurobiological lines of inquiry into the mind-body problem consists of attempts to discover the neural correlates of consciousness (NCC).[19] This entails identifying the minimal set of neuronal events and mechanisms jointly sufficient for a specific conscious experience. According to the latest findings, the NCC are believed to consist of synchronous firing activity of neurons in the forebrain, involving sequences of pulses, about a tenth of a volt in amplitude and 0.5 to 1 milliseconds (msec) in duration.[20] This has yet to be proven. But by itself, this definition of the NCC does not provide a complete explanation for the emergence or nature of mental phenomena.

The fact that a set of *neuronal* processes is necessary for generating a specific mental experience does not preclude the possibility of other *non-neuronal* factors being necessary for generating that experience. Furthermore, the identification of NCC that are *necessary* causes for a specific conscious experience in humans does not imply that they are either *necessary or sufficient* for generating a similar mental event in all other primates, let alone in more primitive animals or even plants. The discovery of the NCC for a specific conscious event may be relatively straightforward when that event is expressed in behavior or is verbally reported. But determining the NCC is highly problematic when that is not the case, as with a human embryo or a person in a vegetative state. Moreover, while it is sometimes said that a person who is asleep is unconscious, we are aware of the contents of our dreams (and possibly of the fact that we are dreaming), and even in dreamless sleep there is a low level of consciousness, including the subliminal ability to discern different kinds of sensory stimulation from the physical environment. For instance, a mother may remain asleep in the midst of loud traffic noise but immediately awaken at the sound of her infant crying. There must presumably be a minimal set of causal neural events for this subtle degree of consciousness. Identifying them, however, is difficult, because scientists cannot cross-reference them with any behavioral expressions until the subject has returned to waking consciousness.

Even if neurobiologists eventually identify a minimal set of physical events that is *sufficient* for a specific conscious process, if the NCC *precede* that mental event (commonly by about 100 msec), this implies a *causal relationship* between a prior neuronal activity and a subsequent mental event. But then the NCC cannot be *identical* to their correlated mental effects, for the two do not exist at the same time. While it may eventually be possible to identify NCC that occur at the same time as a specific mental process, that will still not prove that the two are equivalent. *At present there is no empirical evidence that any neural process is equivalent to any mental process*, and it is far from clear how to go about determining such equivalence. Neuroscientists can precisely measure the time at which a neuronal event occurs, but it is far more challenging to establish the exact moment a subjective experience takes place. The latter can be observed only from a first-person perspective, and there is always a lag between the experience and its verbal report or behavioral expressions, which may vary from one person to the next.

If one categorically insists that mental phenomena must be objectively and physically measurable in order to be deemed existent, then the NCC will be the most obvious candidates for defining consciousness in the language of physics and biology. The only thing left out will be the characteristics of mental phenomena as they actually appear and are experienced from a first-person perspective. This glaring omission has led some neuroscientists and philosophers to acknowledge that mental phenomena cannot be completely understood in terms of their neural correlates, even though subjective experience is physical in nature.[21] According to one theory, certain neural processes have both an objective physical aspect (which is real) and a subjective feel that appears to be nonphysical (which is illusory).[22] This approach is not without precedent in the history of psychology. Behaviorists, who also refused to accept the *appearances* of mental processes at face value, were intent on identifying their *essential nature* in terms of *behavioral dispositions* for responding to stimuli. In this more recent attempt at a physicalist interpretation of the mind, a similar distinction is drawn between the *first-person experience* of mental events and their *physical realization* in the brain. But, as pointed out in the preceding chapter, no one has been able to explain what it is about certain (as yet unidentified) neural processes that enables them to take on this mysterious dual nature—objective and subjective—and to thereby "realize" mental phenomena. This is simply a reformulation of the so-called hard problem, for which no compelling solution has been devised.[23]

## The Natural World = The World of Physical Causality

One of the problems of the dual-nature theory of the neural correlates of consciousness is that it appears to be unfalsifiable, which casts doubt on its status as a scientific theory rather than a mere expression of an ideological commitment to neo-Darwinism. Its advocates would counter, however, that there is an enormous body of indirect evidence suggesting that if subjectively experienced mental processes causally influence human behavior, they must be physical—for only physical phenomena can causally interact with other physical phenomena. With this in mind, neurobiologists seek the "neural mechanisms" by which mental phenomena are realized and by which subjective experience may influence the brain and behavior. While this is a very worthwhile line of inquiry, the assumption that all natural phenomena causally interact only by way of physical mechanisms was undermined in 1887 by the Michelson-Morley experiment, which showed that there is no mechanical explanation for the propagation of electromagnetic fields through empty space. The belief that all causally effective natural phenomena consist of matter and its functions was undermined in 1915 by the general theory of relativity, which shows that there are reciprocal causal interactions between space-time and configurations of mass-energy. And the belief that all causally effective natural phenomena consist of configurations of mass-energy and space-time has been undermined by quantum field theory, which reduces all forms of mass-energy to oscillations of immaterial, mathematical quantities in empty space.

Despite these advances, many cognitive scientists (whose professional training does not require the study of modern physics) still insist that our only choices of theories regarding the mind-body problem are Cartesian dualism or materialistic monism. In light of modern psychology, Cartesian dualism is no longer a viable option, and in light of modern physics, materialistic monism is just as antiquated. Contemporary physics presents at least three classes of phenomena that are causally effective in the natural world: mass-energy, space-time, and abstract, mathematical quantities, such as principles of symmetry. So science has gone beyond both monism and dualism to a realm of empirical pluralism. Presenting Cartesian dualism and materialistic monism as the only two options for understanding the nature of consciousness is like presenting two impossible dishes on a menu, passenger pigeon breast and marinated dodo: they're equally extinct, so neither is a real option in today's world.

While the neural correlates of consciousness undoubtedly influence mental phenomena, do conscious mental states and processes reciprocally influence the brain? While many neuroscientists still believe that the mind is passive—implying that the brain takes on the role of an idol in relation to mental phenomena—as noted earlier, physicists have yet to discover any natural relationship between two phenomena in which only one is influenced by the other. Under the usual conditions of experimental physics or biology, the influence of consciousness may appear very small, but the same may be said of the relation of light to mechanical objects. Mechanical objects influence light—otherwise we could not see them—but experiments to demonstrate the effect of light on the motion of mechanical bodies are difficult. Such effects were first suggested by theoretical considerations concerning the phenomenon of light pressure, and since the mid-twentieth century, they have been amply demonstrated with the use of lasers. The causal efficacy of the mind is already being scientifically studied as it manifests in placebo effects of all kinds and in the effects of mental training on the brain. It remains to be seen whether consciousness may be focused in a laserlike fashion, and whether such high-energy consciousness might require scientists to modify the laws of physics.[24]

Despite the many unresolved questions concerning the nature and origins of consciousness, the great majority of cognitive scientists and philosophers today express confidence that there is a simple solution to the mind-body problem, and that it has been available to any educated person since serious research on the brain began nearly a century ago: mental phenomena are caused by neurophysiological processes in the brain and are themselves features of the brain.[25] But before this hypothetical solution can be validated, the neural correlates of consciousness must be identified, and tests must be conducted to determine whether the NCC are both necessary and sufficient for the experience of consciousness. To establish that they are causally *necessary*, scientists have to find out whether a subject who has the alleged NCC removed thereby loses consciousness. To establish that they are causally *sufficient*, scientists have to discover whether an otherwise unconscious subject can be brought to consciousness by inducing the alleged NCC. No one has yet accomplished the first step, identifying the NCC.[26]

At present, cognitive scientists do not have the slightest idea how anything material could be conscious, but because of their commitment to naturalism of one sort or another, they conceal their ignorance of the nature and origins of consciousness with illusions of knowledge, based on future discoveries they hope will be made to validate their present beliefs. But this approach is contrary to the scientific attitude, which Erwin Schrödinger summed up: "Instead of filling a gap by guesswork, genuine science prefers to put up with it."[27]

Christof Koch, one of the leading researchers investigating the neural correlates of consciousness, points out that according to physicalism, neuronal and mental events are identical: the neural correlate for a specific mental process *is* that subjective experience. While the former is measured by microelectrodes, the latter is experienced by brains, and the former is sufficient for the latter. But even he, who has long advocated a staunch materialist view of the mind-body problem, has begun to express doubts: "Are they really one and the same thing, viewed from different perspectives? The characters of brain states and of phenomenal states appear too different to be completely reducible to each other. I suspect that the relationship is more complex than traditionally envisioned. For now, it is best to keep an open mind on this matter and to concentrate on identifying the correlates of consciousness in the brain."[28]

As long as cognitive scientists continue to investigate the mind primarily by way of its physical correlates—behavioral expressions and neural causes—they have no way of testing their physicalist assumptions about the nature of mental phenomena. This very mode of inquiry *assumes* their equivalence with their neural correlates, without offering any means of verifying that equivalence. Since science has historically been equated with objective science, it has understandably, but unjustifiably, omitted consciousness and all subjective mental phenomena from the natural world. It is time now to naturalize the mind and include mental phenomena among the growing list of natural phenomena—together with mass-energy, space-time, and mathematical quantities—that are real and causally efficacious.

The history of naturalism in modern science has followed the same trend as the earlier history of theism in science. Virtually all the great pioneers of the scientific revolution were theists, with a robust belief in the nature of God and his creative and regulative role in the natural world. But with the advances in physics in the eighteenth century, a growing number of scientists downgraded their religious convictions from theism to deism, denuding God of many of his earlier qualities and leaving him a relatively passive role after his initial creation of the universe. And with the advances in geology and biology in the nineteenth century, deism gradually began to give way to agnosticism. Toward the end of that century, Friedrich Nietzsche summed up this movement with his famous declaration in *Also sprach Zarathustra*: "God is dead. God remains dead. And we have killed him. How shall we, murderers of all murderers, console ourselves?"

The second revolution in physics in the twentieth century began to undermine the very notion of matter. So informed naturalists have been forced to retreat from their robust belief in the existence of matter and its role in creating and regulating the natural world. A growing number of staunch eliminative materialists, who deny the very existence of subjective mental states, are abandoning their earlier views and joining the ranks of "nonreductive physicalists," who acknowledge the existence and causal efficacy of mental states, which are "realized" as physical states. This reflects a pattern much like the descent from theism to deism. If the past is any key to the future, we may expect that the erosion from theism to deism to agnosticism will be reflected in a progression from materialism to physicalism to agnosticism. This will culminate in the frank acknowledgment that scientists do not know what consciousness is, how to measure it, what causes it, or what role it has in nature. All illusions of knowledge of religion and science will be abandoned, and that may open the way for a true revolution in the mind sciences.

# 3

# TOWARD A NATURAL THEORY OF HUMAN CONSCIOUSNESS

AS DISCUSSED in the previous chapter, most of the current naturalistic accounts of consciousness that have been devised by cognitive scientists and philosophers are based on a materialistic view of the universe that was prevalent in the late nineteenth century, when the scientific study of the mind began. A central premise of this book is that the lack of a major revolution in the cognitive sciences is due in part to the antiquated notions of physics that underlie most contemporary theorizing about the nature of consciousness. In seeking to understand the role of the mind in nature, psychologists rely on biologists and biologists rely on physicists. In this chapter I shall discuss some of the most provocative ideas about consciousness presented by leading physicists, with the goal of facilitating a greater degree of dialogue and perhaps collaborative research between contemporary physicists and cognitive scientists, who normally work in isolation from each other.

Nowadays there are two schools of thought about the significance of quantum theory for understanding the world of nature. The more prevalent school, which we may call the exclusivists, says that quantum mechanics covers only a small part of physics, namely the part with events on a local or limited scale. The major historical exponent of this view was Niels Bohr, who maintained that quantum mechanics can describe only processes occurring within a larger framework that must be defined classically. Most cognitive scientists and philosophers of mind today assume this to be true. The other school, which I call the inclusivists, declares that quantum mechanics applies to all physical processes equally. The leading exponent of this view is Stephen Hawking, who is trying to create a theory of quantum cosmology with a single wave function for the whole universe.

Bohr's understanding of physics is based on the principle of complementarity, which says that nature is too subtle to be described adequately by any single viewpoint. In his perspective, classical physics deals with facts and quantum physics is concerned with probabilities. In his fusion of the two, the world consists of an inseparable mixture of probabilities and facts, so our description of it must likewise be an inseparable mixture of quantum and classical explanations. A fundamental problem with this dualistic view of reality is finding any objective criterion for demarcating quantum and classical, since mathematical abstractions (in the quantum world) somehow turn into concrete realities (in the classical world).

This problem is avoided by quantum cosmologists, who insist that the quantum picture must include everything and explain everything. According to the mainstream view, the classical picture must be built out of the quantum picture by a process called decoherence. This is the interactive process of a quantum system with the macroscopic environment, in which wave-interference effects seen in quantum systems are very rapidly dissipated. This results in a set of classical outcomes, each having its own probability of being realized. The theory of decoherence also provides a justification in the minds of many theorists today for discounting quantum mechanics when attempting to understand mind-brain interactions. But decoherence has its own problems, which physicists continue to grapple with.[1]

Physicist Freeman J. Dyson points out that there are two kinds of science, known to historians as Baconian and Cartesian. Francis Bacon primarily emphasized the careful examination of natural phenomena, without overinterpreting them or obscuring them with beliefs and preconceptions. René Descartes, on the other hand, formulated his laws of nature primarily on the basis of his belief in the infinite perfection of God, so that even if God had created many worlds, all of those laws would necessarily be observed in each one. Modern science has evolved by way of the dynamic competition between Baconian and Cartesian viewpoints, which have proven to have "complementarity," in the sense that Niels Bohr used this term. Both have validity and both are necessary for the further evolution of science, but they cannot be seen simultaneously. As Dyson comments, "We need Baconian scientists to explore the universe and find out what is there to be explained. We need Cartesian scientists to explain and unify what we have found."[2] At this point in history, the scientific study of the mind is in dire need of Baconian scientists who are committed to the rigorous observation of mental phenomena, so that Cartesian scientists can explain those empirical discoveries and unify them within the larger framework of contemporary science. But in this secular age, Cartesian scientists no longer invoke the perfection of God to explain the orderly world. Instead they invoke the perfection of the principles of scientific materialism, which are firmly rooted in nineteenth-century classical physics.[3]

Among theoretical physicists over the past few decades, no one has proposed more innovative and revolutionary hypotheses concerning the role of consciousness in the natural world than John Archibald Wheeler. Central to his thinking was an integration of the diverse domains of cosmology and atomic physics. He speculated that the role of the observer is crucial to the laws of physics, not only at those two extremes, where it has thus far been noticeable, but also over the whole range in between. A complete determination of the laws of physics, he maintains, must include a compelling account of the role of the observer.

Anton Zeilinger, a leading experimental physicist in the foundations of quantum mechanics, argues, "The outstanding feature of Professor Wheeler's viewpoint is his realization that the implications of quantum mechanics are so far-reaching that they require a completely novel approach in our view of reality and in the way we see our role in the universe. This distinguishes him from many others who in one way or another tried to save pre-quantum viewpoints, particularly the obviously wrong notion of a reality independent of us."[4] While many theorists assume that quantum effects occur only at very low temperatures, Zeilinger points out that quantum interference patterns can be observed at 900° Kelvin. In principle, he says, nothing in quantum physics limits the size of objects for which such interference might be observed someday.

Quantum physicists often refer to elementary particles being in a superposition state: in neither one place nor another, but ambiguously in both places at once. As long as that superposition remains, there is no way to tell whether the particle is here or there. Physicists have to leave the ambiguity open. Since no size limit has been found experimentally for the validity of quantum superpositions, experimentalists have only to further develop and refine their techniques in order to extend the realm of systems for which quantum interference has effects to larger and larger systems, perhaps including living organisms.

A major stumbling block in considering the dynamic role of mental phenomena in mind-brain interactions is the Cartesian insistence on providing mechanical explanations for all kinds of causal influences. Some philosophers propose "top-down" mechanical influences of the mind on the body, equating mental phenomena with "higher-order" neural functions. But providing a mechanical explanation before empirical evidence of causality can be admitted is at variance with the history of science. Newton's laws of motion, including those pertaining to gravitation, were based on decades of rigorous observations of physical phenomena, but he was unable to provide a mechanical explanation for the force of gravity through empty space. This took 228 years, from 1687, when he published his major work,[5] until 1915, when Einstein published his general theory of relativity, which explained gravity in terms of the curvature of space-time. In other domains of physics, the mechanical explanations for well-known causal interactions have simply been discarded on both theoretical and empirical grounds. As mentioned earlier, since 1887, there has been no mechanical explanation for the propagation of electromagnetic fields through empty space, and quantum theory has undermined previously assumed mechanical explanations of the causal interactions among elementary particles and fields. Moreover, when a quantum measurement takes place, there is no physical mechanism by which probabilities turn into actualities, no mechanism by which a real world emerges from a world of potentialities.

In the life sciences, Darwin's theory of natural selection was based on decades of meticulous observations, but a hundred years passed between the publication of *The Origin of Species*[6] in 1859 and Francis Crick and James Watson's discovery of the DNA molecule, which opened the way to understanding the biological mechanisms that make natural selection possible. No one knows how objective brain processes generate or even influence subjective experience, or how mental phenomena influence the brain and behavior. To balance out the present Cartesian insistence on mechanical explanations for mind-body interactions, the scientific study of the mind needs a healthy dose of Baconian empiricism to examine the appearances of such interactions with as few preconceptions as possible.

Regarding the role of consciousness in nature, Stanford physicist Andrei Linde suggests that scientists occasionally allow themselves to overcome their natural conservatism regarding theories that appear "metaphysical," and take the risk of abandoning some of their standard assumptions. One of these is that consciousness, just like space-time before the theory of general relativity, plays a secondary, subservient role in the universe, as nothing more than a function of matter. This scientific view of nature as matter obeying laws of physics is so successful that we easily forget that everything we know of the objective world is by way of human consciousness. The objective world of matter becomes the only reality, taking on the role of an idol in its causal interactions with subjective states of consciousness. This assumption is almost as natural and perhaps as false, he says, as our previous assumption that space is only a mathematical tool for the description of matter. We are substituting a working theory of an independently existing material world for the reality of first-person experience, and that theory works so well in many regards that we almost never think about its possible limitations.

Continuing this provocative line of thought, Linde asks, "Is it possible that consciousness, like space-time, has its own intrinsic degrees of freedom, and that neglecting these will lead to a description of the universe that is fundamentally incomplete? What if our perceptions are as real [as] (or maybe, in a certain sense, are even more real) than material objects?"[7] Nothing in physics prevents us from adding a "space of elements of consciousness" to the natural world consisting of mass-energy, space-time, and informational states. This hypothesis would open the way to investigating the possibility that consciousness may exist by itself, even in the absence of matter, just like gravitational waves, excitations of space, may exist in the absence of protons and electrons. Exploring this parallel further, he points out that gravitational waves usually are so small and interact with matter so weakly that empirical evidence for them has yet to be found. However, their existence is absolutely crucial for understanding certain astronomical data. Perhaps consciousness plays an equally important role in nature, despite the fact that it has been ignored until now in understanding well-studied physical processes in the brain and elsewhere. As we restore the balance between Baconian science and Cartesian science, Linde suggests that we may find the study of the universe and the study of consciousness inseparably linked, so that ultimate progress in the one will be impossible without progress in the other.

This general hypothesis brings us back to the significance of quantum mechanics in the universe. All matter was produced by quantum processes after the end of inflation, the extremely rapid expansion of space-time following the big bang. All galaxies were produced by quantum fluctuations generated in the last stages of inflation. As one of the leading experts in quantum cosmology, Linde summarizes these findings with the statement: "*Without inflation, our universe would be ugly. Without quantum, our universe would be empty.*"[8] Some astrophysicists are currently proposing inflationary scenarios that include an infinite number of "pocket" universes exploding from fluctuations within the quantum vacuum into myriad and diverse cosmoses. Although these other worlds cannot be observed directly, on the basis of empirical data provided by the Cosmic Background Explorer satellite, these scientists have produced many indirect arguments in support of the inflationary model of the very early universe.

Michael B. Mensky, a physicist at the Lebedev Physical Institute of the Academy of Science in Moscow, further explores the interface between quantum theory and the scientific understanding of consciousness. The theory of decoherence adequately explains the existence of different alternative results of measurement in quantum mechanics, each with its own probability, but he is intent on understanding the process by which any one of these alternatives is selected. According to a metaphor proposed by Wheeler, at the time of quantum measurement the observer faces a kind of railway switch that regulates which of several directions his train will follow. Depending on the direction determined by the switch, the observer will see one or another result of measurement. The possible directions correspond to the alternative results of quantum measurements. Mensky concludes that *"a theory that would describe not only the set of alternative results of measurement and the associated probability distribution, but also the mechanism of selection of one particular result, ought to include the mind (consciousness) of the observer."*[9]

This leaves two unresolved problems: the selection of one alternative in quantum measurement and the role of consciousness in the natural world. There are cases in the history of science, he notes, when two formidable problems were solved simultaneously, as though helping to solve each other. And he suggests that the measurement problem in quantum mechanics and the problem of consciousness in cognitive science constitute such a pair of deeply interconnected issues.

In a review of Mensky's speculations regarding the relevance of consciousness for solving the measurement problem, Vitaly L. Ginzburg, co-recipient of the 2003 Nobel Prize in physics, begins by acknowledging that scientists have not satisfactorily explained the origin of life and consciousness, so it would be a mistake to categorically dismiss Mensky's discussion of the origin of human consciousness and its relation to quantum mechanics. Such informed theorizing is precisely what is required, he adds, because progress in the interpretation of the quantum theory of measurement is impossible without further analysis.[10] Providing a broader context for this kind of interdisciplinary inquiry, Ginzburg asserts that two of the most important and interesting problems in physics at the beginning of the twenty-first century are the interpretation of quantum mechanics and the problem of reductionism, that is, the question of whether the phenomenon of life can be explained on the basis of presently known physics.[11]

Mensky focuses primarily on the first of these, arguing that "the immanent feature of quantum mechanics (more precisely, of quantum physics, including relativistic physics) that distinguishes it from all remaining physics is that attempts to represent the measurement process in it as completely objective, as absolutely independent of the observer who perceives the result of the measurement, have not met with success."[12] Moreover, the progress of quantum mechanics in the last two to three decades has made the inclusion of the observer's consciousness an absolute necessity when discussing conceptual problems. Until now, most physicists have sought to describe the properties of the material substance that generates consciousness, namely, the brain or some structure inside it. Researchers have relied on the decoherence theory, but it has become increasingly clear that this is not the answer. Consequently, consciousness has increasingly come to be regarded as a natural something that can be phenomenologically described but cannot be derived from the known properties of (quantum) matter. Mensky regards the role of consciousness with respect to the measurement problem as the "problem of the century," and the way forward, he maintains, is to search first for conceptual solutions rather than mathematical ones.

Some experimental physicists have taken up the challenge of investigating the effects of consciousness on the brain in light of quantum field theory, the most fundamental theory of physics at both the microscopic scale and the macroscopic scale. Beginning in the 1960s, Hiroomi Umezawa, a Japanese Nobel laureate in physics, developed a framework of quantum field theory to describe fundamental processes in macroscopic living matter. In the 1970s, with the help of another Japanese physicist, Yoshiyuki Takahashi, he went on to develop a standard field theoretical model of the memory mechanism in the brain. Then in the 1990s, this Umezawa-Takahashi model was formalized by Mari Jibu and her colleagues into a concrete theory related to consciousness called "quantum brain dynamics."[13] A sophisticated version has also been developed by Charles Enz, the last collaborator of Wolfgang Pauli.[14]

Any suggestion that states of consciousness may exist by themselves, independently of matter, and have causal efficacy in the physical world is bound to raise concerns about such an immaterial space of consciousness violating the well-known principle of the conservation of mass-energy. John Wheeler has likened the history of physics to a staircase of transcendence, at each step of which some assumed physical property has been replaced by a new conceptual scheme. For example, the discovery of nuclear transmutations showed that the law of conservation of the elements could be transcended. And when a massive star collapses, the gravitational field enlarges to such an extent that even light itself is trapped. At that time, the material of the core of the star retreats inside a so-called event horizon and effectively disappears as far as the outside universe is concerned. Theory suggests that only a handful of parameters survive the collapse, with mass, electric charge, and angular momentum being the three principal conserved quantities. Otherwise, cherished conservation laws are not so much violated as transcended: they cease to be relevant.

According to Wheeler, the lesson to be learned from this evolution of physics is that "Law cannot stand engraved on a tablet or stone for all eternity. . . . All is mutable."[15] Physicist Paul C.W. Davies comments, "In this respect Wheeler was breaking a 400-year-old scientific tradition of regarding nature as subject to eternal laws. Second, the very appearance of lawlike behavior in nature might be linked in some way to our observations of nature—subject and object, observer and observed, interwoven. These were radical ideas indeed."[16]

The Heisenberg Uncertainty Principle suggests that violations of the principle of energy conservation can occur due to spontaneous, unpredictable fluctuations of the vacuum that is space. This has been validated by innumerable experiments. According to quantum mechanics, energy may surge out of nowhere for a brief moment; the shorter the interval, the bigger the energy excursion. When electromagnetic waves pass through space, they produce oscillations at every possible frequency, and when you add up all those ceaseless fluctuations, you get a background sea of light whose total energy is enormous. This is called the zero-point field of empty space. The "zero-point" energy of this field is huge, even though it is the field's lowest possible energy state, and all other energy in space is over and above it.

Since the zero-point field is everywhere, we are effectively blind to it, while the world of light that we do see is all the rest of the light beyond the zero-point field. A theory that will be developed in the following chapters suggests that the vacuum may be not only filled with zero-point energy, which can be objectively measured with techniques of physics, but also permeated with consciousness, which can be subjectively experienced with techniques of introspection.

While classical physics denies the possibility of causal efficacy by any nonphysical entity, that is, anything not consisting of a configuration of mass-energy, modern physics has shown that space-time and information have a causal role in nature. So a more contemporary version of naturalism acknowledges that there are natural but "nonphysical" properties in the universe, such as "informational states." Physicists remain divided as to what these are. Some define them objectively in terms of entropy, while others, including Wheeler, insist that they must be semantically meaningful, must involve a consciousness that is informed of the content of the information.

In this chapter I have narrated some of the most provocative ideas about consciousness and its role in nature expressed by leading physicists on the cutting edge of theoretical and empirical research—not to compel agreement, but to provoke further theorizing leading to empirical research. For, as John Wheeler noted, "Progress in science owes more to the clash of ideas than the steady accumulation of facts."[17] In the true spirit of Baconian science, I turn now to a line of empirical inquiry that may shed fresh light on the phenomenon of consciousness.

# 4

# OBSERVING THE SPACE OF THE MIND

## Parallels in Astronomy

For thousands of years, people have been fascinated by the night sky and observed celestial phenomena very carefully, but with the unaided eye, only a few thousand stars can be seen. Everything else remained hidden in the "subconscious" of deep space, beyond the scope of empirical research and therefore confined to the domain of metaphysics until 1609, when Galileo heard of the telescope invented by a Flemish spectacle maker, Hans Lipperhey, and swiftly constructed one for himself. His first attempt produced an eight-power telescope, which he later increased to twenty-power by grinding his own lenses, and he used his new instruments for observing the heavens in ways never before attempted. The next year he published his findings in a book, *The Starry Messenger*, in which he reported not only his observations of the moons of Jupiter but also his discovery that the Milky Way consists of a vast collection of stars that had never been seen before. In this way, the depths of the physical universe previously concealed from human consciousness began to be explored.

The science of astronomy has continuously progressed since Galileo's time, but it was more than 300 years before scientists discovered galaxies beyond the Milky Way. As a result of a series of observations in 1923–1924, the American astronomer Edwin Powell Hubble, using the newly completed 100-inch Hooker Telescope at Mount Wilson, established beyond doubt that the fuzzy "nebulae" seen earlier with less powerful telescopes were not part of our galaxy, as had been thought, but galaxies themselves, outside the Milky Way. Hubble announced his discovery in 1924, and five years later, together with another American astronomer, Milton Humason, he formulated the empirical Redshift Distance Law of galaxies, or "Hubble's law," which states that the greater the distance between any two galaxies, the greater their relative speed of separation. This influenced the formulation of the big bang theory by George Gamow in 1948, for which the discovery of cosmic background radiation in 1965 provided empirical support.

Some of the most recent probes into deep space, made with the Hubble Space Telescope in 2003–2004, have unveiled the most detailed portrait of the visible universe ever achieved by humankind. The Hubble Ultra Deep Field, a million-second-long photo exposure taken over the course of 400 Hubble orbits around Earth, reveals the first galaxies to emerge from the so-called "dark ages," the time shortly after the big bang when the first stars reheated the cold, dark universe. The telescope was directed to a region of space in the constellation Fornax, of which ground-based telescopic images appear mostly empty. But in this long exposure from the orbiting Hubble telescope, with photons from the very faintest objects in space arriving at a trickle of one photon per minute, scientists were able to acquire a "deep" core sample of the universe, cutting across billions of light-years. By peering into a patch of sky just one-tenth the diameter of the full moon, scientists brought into view nearly 10,000 galaxies, some of them existing when the universe was only 800 million years old. The whole sky contains 12.7 million times more area than this Ultra Deep Field. Scientists expect that such observations will offer new insights into the birth and evolution of galaxies.

This brief history of astronomy gives some idea of the importance of sophisticated, penetrating observation for exploring the depths of space and the evolution of the physical universe. But such objective observations tell us nothing about the role of the observer in relation to the quantum fluctuations in the last stages of inflation after the big bang, without which there would be no galaxies and no matter in our universe.

## Philosophical Resistance to Introspection

As discussed in the first chapter, since the time of Descartes, scientists have taken on the challenge of exploring the world of objective physical phenomena, leaving the world of subjective mental phenomena to philosophers. Renaissance philosophers such as Paracelsus, who advocated an organic philosophy in contrast to the mechanistic philosophy of Descartes, did emphasize the first-person observation of the mind and first-person experimentation using the power of imagination (*vis imaginativa*). But they lived in the tragically psychotic era of witch hunting, during which any such notions were suspiciously regarded as magic. Protestant reformers were especially quick to condemn anything of that sort as impious, useless, and potentially demonic, and those who advocated such theories and methods could find their lives imperiled. In contrast, Bacon's empiricism, which was confined to the objective world, was perfectly consistent with the new Protestant work ethic and the prevalent fear of probing the depths of the human psyche.

Since that time, instead of developing rigorous means to experientially explore the subjective dimensions of the natural world, generations of philosophers have devised ingenious arguments for denying that the mind can be explored from a first-person perspective. Immanuel Kant, for instance, claimed that due to the subjective nature of mental phenomena, any introspective observations could at most provide a *historical* account, not a true, "objective" *science*. But if "real-time" observations were a requirement for any objective science, the whole of astronomy would fail to meet it. Even observations of the moon entail a time lag of more than a second, observations of the sun and planets record events minutes after they have taken place, and our knowledge of distant galaxies is billions of years old. Due to delays caused by the speed of light, astronomers may be regarded as "celestial journalists" with regard to the solar system and "historians" with regard to their observations of the rest of the universe. In twenty-first-century astronomy, historical accounts of the universe are the most we can ever hope for. In the introspective study of the mind, there are certainly many mental phenomena, such as emotions, that may be "observed" only retrospectively by way of memory. But there are many other mental phenomena, such as mental chit-chat, deliberately induced mental images, and dreamscapes, that are observed in real time. Arguably, the introspective observations of mental events as they occur are the only truly "real-time" accounts available to us. For even the visual and auditory perceptions of nearby colors and sounds are slightly delayed due to the speeds of light and sound.

Kant further argued that there could be no true science of the mind based on introspection since the observed mental phenomena are altered and transformed by the very act of observation.[1] Niels Bohr was among the first physicists to note the observer participancy parallel between examining mental phenomena and examining quantum processes. In quantum measurement, the act of observation invariably alters the observed phenomena, but that has not prevented quantum mechanics from becoming the most successful physical theory in the history of science.

In many experiments, it has been demonstrated that objects do not exist in a well-defined way prior to the act of measurement. For example, when single photons are emitted by a source so low in intensity that the probability of the simultaneous arrival of more than one photon at the detector is negligible, it is possible to count the number of detector actuations and thereby find the number of arriving photons. But it turns out that a light field cannot be represented as a collection of a definite number of photons, for the number of photons in it is not defined prior to the instant of measurement![2]

Moreover, the extent to which mental events are altered and transformed by the very act of observation is variable. One testable hypothesis is that with training, one may observe mental phenomena more and more "objectively," so as to exert less and less influence on what is being observed. This may occur in the dream state as well as the waking state. For example, one may observe events in a lucid dream (in which dreamers are aware that they are dreaming) without overtly altering them. Of course, there is still observer participancy, so the comparison with quantum mechanics is an excellent one, but in neither case does this imply that the objects being observed are mere artifacts of the method of observation.

Among cognitive scientists, William James took the bold step of emphasizing the primacy of introspection for the scientific study of the mind,[3] and among philosophers, Edmund Husserl made a worthy attempt at developing a phenomenology of consciousness with his method of "bracketing" consciousness from its object.[4] But twentieth-century philosophers have continued to raise serious questions about the possibility, let alone the efficacy, of developing a science of the mind based on the direct observation of mental phenomena.

Ludwig Wittgenstein, for instance, divided mental vocabulary into two classes: world-directed concepts and mind-directed concepts. Regarding the latter, he challenged the very possibility of a "private language" conveying meaningful information about internal experiences of being conscious.[5] In support of this argument, it is true that science requires theories to be intersubjectively re-testable by replicating experiments with suitable instruments. But in addition, to test any sophisticated theory, the experimenters must have professional training in the use of those instruments and in interpreting the data produced. In modern scientific research, an untrained person called in from the street rarely qualifies as a suitable "third person" who can either validate or invalidate a previous finding.

A crucial element of scientific inquiry since the time of Pythagoras has been mathematics, which has taken on an especially prominent role since the scientific revolution. In 1623 Galileo famously wrote: "Philosophy is written in this grand book—the universe—which stands continuously open to our gaze. But the book cannot be understood unless one first learns to comprehend the language and interpret the characters in which it is written. It is written in the language of mathematics, and its characters are triangles, circles, and other geometrical figures, without which it is humanly impossible to understand a single word of it; without these one is wandering about in a dark labyrinth."[6] And the practice of higher mathematics takes place within the mind of the mathematician and is then communicated to other mathematicians. Writing equations on a chalkboard is simply a kind of public behavior that may or may not result from the internal process of understanding proofs and devising theorems. A mathematically uneducated person may be taught how to write down the same equations, but when subjected to interrogation by a qualified mathematician, will clearly not understand what he has written. Mathematicians do commonly converse among themselves in a kind of language that is unintelligible to nonmathematicians, and the same is true of experts in all fields of science. So there is no reason in principle that researchers could not receive professional training in observing mental phenomena and learn to communicate among themselves about their experiences. However, this is a major undertaking that neither philosophers nor cognitive scientists have yet tackled.

Sigmund Freud raised a formidable practical concern about the prospects for making unbiased observations of one's own mind: there are conscious and unconscious impulses in the mind that may sometimes conceal thoughts, memories, emotions, and desires we would prefer not to acknowledge, and we may imagine such mental processes even though they are not present.[7] Albert Einstein is credited with the statement, "Only two things are infinite, the universe and human stupidity, and I'm not sure about the former," and this is a serious concern for raising introspection to a scientifically rigorous status. What is required is relentless self-honesty, which may be cultivated with intensive, prolonged training in introspection. This is where the validity of introspective observations may be cross-checked with sophisticated ways of evaluating behavior and determining the neural correlates of mental states and processes. This threefold approach is precisely what William James advocated when he set forth his strategy for the scientific study of the mind.

Psychologists have a lot of evidence to show that perception is a function of expectation, and introspective perception is clearly not immune to such influences.[8] Both sensory and introspective experiences are precognitively structured; those structures enable us to perceive things in terms of specific aspects; and those aspects are constrained by our familiarity with sets of categories that enable us, in varying degrees, to assimilate our experiences, however novel, to the familiar. Making genuine discoveries in the space of the mind by means of introspection will evidently require months or years of rigorous training, and once again, cross-checking findings with behavioral and neural analyses.

Subtle distinctions must also be made, for example, between *imagining* that one desires something and *actually* desiring it. Within the space of the mind, superficial appearances do not always correspond to reality, especially when they have been sifted through complex and often subliminal processes of interpretation. In addition to this pragmatic psychological question, Gilbert Ryle raises the philosophical concern about making ontological inferences about the way mind *is* from the way mental states *seem*.[9] This relates to an issue discussed in the first chapter: the fact that mental phenomena appear to bear no distinctively physical attributes at all. But if one assumes that everything that exists must be physical, then the appearances of mental phenomena must be illusory.

This is precisely where the Baconian and Cartesian approaches to scientific inquiry diverge. If we follow Bacon's emphasis on empirical induction and apply it (as he did not) to the examination of subjective experience, we will be inclined to learn as much as possible about the mind by observing mental phenomena themselves. But if we follow Descartes' deductive, rationalistic lead as it has been adapted by scientific materialists, then we will focus almost entirely on the physical correlates of consciousness, while marginalizing the observation of mental phenomena. Evidently, mainstream philosophy, psychology, and neuroscience have embraced the latter option. There are strengths and weaknesses to this approach; I am suggesting that it may be well complemented, not supplanted, by the incorporation of refined introspection into the scientific study of the mind.[10]

## Developing a Telescope for the Mind

Philosophers have been debating the merits, limitations, and defects of introspection for centuries, but they do not seem to have refined our capacity for observing mental phenomena. We are as far as we ever were from developing a telescope for the mind. A thesis can in principle be proved or strongly argued, whereas a stance—such as a particular approach to scientific inquiry—can be adopted only by a sort of "Gestalt-switch." And this is what I am proposing: a Gestalt-switch *away* from the common tendency to empirically and theoretically marginalize introspection *to* accepting the formidable challenge of enhancing introspection in ways that are unprecedented in the history of modern science. This implies a return to empiricism: taking the *methods* for making penetrating observations of all kinds of natural phenomena to be of the highest value, instead of assuming that the materialist *ideology* in its present formulation already provides a key to unlocking all the remaining mysteries of nature.

Scientific empiricists since Francis Bacon have generally confined their stance to observations of objective physical phenomena, whereas contemplative empiricists claim to have developed their faculty of mental perception to observe the space of the mind. To someone who has not utilized or refined this faculty, which the ancient Greeks called *noētos*, contemplatives' experiential reports may sound like nothing more than speculation. The semiprivate language of highly trained contemplatives, like that of professional mathematicians, therefore becomes either unintelligible to or misinterpreted by laypeople.

Over the past three millennia, contemplative traditions of varying degrees of sophistication have developed in the East and West, and one point on which they all seem to agree is the need to refine one's attention skills in order to make reliable observations of mental phenomena. Specifically, the deeply habituated tendencies of mental agitation and dullness need to be overcome through the development of attentional stability and vividness. These skills may be strengthened in a separate set of mental exercises[11] or in the very process of learning how to observe the mind. Both approaches have been explored in the Hindu, Buddhist, and Taoist traditions of India, Southeast Asia, East Asia, and the Himalayan plateau. In the spirit of healthy, open-minded, scientific skepticism, the alleged discoveries of contemplatives in these traditions should be treated with the same attitude with which scientists respond to any other claim of discovery: see if you can replicate their findings in your own laboratory.

For a minute fraction of the expense of building, maintaining, and operating the Hubble Space Telescope, contemplative observatories could be created for empirical research into the trainability of attention and the possibility of observing the space of the mind with scientific rigor and replicability. Such laboratories would ideally include facilities for conducting behavioral and neuroscientific research, together with simple, individual accommodations for people to devote themselves to mental training for months and years on end. This would be tantamount to creating a new profession of highly trained observers and experimentalists of the mind.

One valuable kind of mental training that I have explained elsewhere entails focusing one's attention on the space of mental events, distinct from appearances generated by the five physical senses.[12] Expertise in this mode of observation may require as much as 5,000 to 10,000 hours of training, 8 to 12 hours a day, 7 days a week, for months on end. In addition to this formal practice of observing the mind and whatever events arise within it, the practitioner must take all necessary steps in terms of lifestyle and emotional regulation to ensure mental health throughout the course of this extremely demanding discipline. Contemplative traditions that have developed such introspective practice have much sound advice to offer in these regards.[13]

As the faculty of mental perception is refined, one may begin probing the nature of the thoughts, images, emotions, and desires that arise in each moment. Specific questions may guide these observations, such as:

- Are any of these mental events, including one's awareness of them, static, or are they constantly in a state of flux?
- Are any mental phenomena inherently satisfying or unsatisfying, or do these qualities arise only relative to one's attitudes and desires?
- Is the space of the mind, any of its contents, or the awareness of them inherently "I" or "mine," or is one's sense of personal identity and possession of one's mind purely a conceptual projection?

## Hypotheses

When a large number of researchers engage in such empirical inquiry in different laboratories, running their experiments with different sets of assumptions and expectations, it may turn out, contrary to Kant's expectations, that they can extract features of the mind independent of the acts of observation. They may be able to identify universal qualities and regularities among mental phenomena and thereby formulate laws of the mind analogous to the rest of the laws of nature. As in any other branch of science, this research will require controlled experiments, repeated iterative evolving cycles of hypothesis formation, controlled testing, hypothesis revision, and prediction.

The above method of observing the space of the mind and everything that arises within it has been practiced in Tibet for more than a thousand years. Those engaged in this practice within a context of religious belief, which certainly colors experience, claim to have made many discoveries that can be replicated by any open-minded individual willing to devote the time and effort to putting their findings to the test.[14] The following discussion highlights some of the alleged discoveries about the mind that may be scientifically treated as hypotheses that can be tested through experience. Such scientific research is already in progress, with one notable project being conducted by the Santa Barbara Institute for Consciousness Studies in collaboration with a team of psychologists and neuroscientists at the University of California, Davis.[15]

With regard to Wittgenstein's concern about the unfeasibility of any private language, Tibetan contemplatives claim that a shared, highly specialized language concerning rarified subjective experience has been developing within a community of professionally trained observers of the mind. Throughout such training, participants converse among themselves and with their mentors and in this way learn to communicate their inner experiences. Nonparticipants overhearing such communication may believe they understand the kinds of experiences being narrated, but in fact most of what is said will be beyond their imagination, for they have never experienced the states of consciousness that are being probed.

Freud's concern about the obscuring and distorting influences of unconscious mental impulses has long been a major concern among Tibetan contemplatives. The remedy they have settled on is relentless, passive but vigilant observation of whatever arises in the space of the mind, without being carried away by or identifying with it. It is imperative not to respond to discursive thoughts, mental images, emotions, and desires with either aversion or craving. Rather, one must simply let them arise and pass of their own accord, without intervening or attempting to suppress or augment them. Metaphorically, one must rest in a "space of awareness" that is larger than the "space of one's own psyche." Whatever arises within the psyche is observed closely and with discerning intelligence, but without modifying, censoring, or editing in any way. This is an extraordinarily demanding endeavor, and it is pursued in close collaboration with an experienced and accomplished mentor who is well versed in such practice.

Buddhist contemplatives throughout Asia have taken special interest in the possible differences between the way mental processes appear and the way they exist, a concern raised more recently in Western research by Gilbert Ryle.[16] Specifically, they have found that although mental states and processes often appear to be relatively static, upon close examination, all the immediate contents of the mind as well as our awareness of them are constantly in flux, arising and passing many times per second. A relatively homogenous continuum of a mental state, such as depression, may endure for seconds or even minutes, but that stream of emotion consists of discrete pulses of awareness, each of finite duration. There is nothing static in the human psyche, though habits may become deeply ingrained over the course of a lifetime.

A second discrepancy between appearances and reality is that certain mental states, such as joy and elation, may appear to be intrinsically satisfying, but upon more careful examination are found to be misleading. No mental state that arises from moment to moment in dependence upon sensory or intellectual stimuli is inherently satisfying. Every affective state is experienced as pleasant, unpleasant, or neutral only in relation to a complex of attitudes and desires. When these affective states of mind are passively observed from the wider perspective of the space of awareness, without identifying with them, they have no absolute, independent attributes of either pleasure or pain.

A third disparity between mental appearances and reality pertains to the fact that thoughts, emotions, and other mental phenomena seem to have an inherent personal quality. When strong identification with these processes occurs, one may feel that one's very identity has become fused with them, and momentarily have the sense "I am angry, " or "I am elated." But with some skill in observing the contents of the mind, one finds that thoughts and mental images arise by themselves, with no voluntary intervention or control by a separate agent or self. Psychophysiological causes and conditions come together to generate these mental events, but there is no evidence that a separate "I" is among those causal influences. To be sure, some thoughts and desires do appear to be under the control of an autonomous self, but as expertise is gained in this practice, this illusion fades away, and everything that arises in the mind is seen to be a natural event, dependent upon impersonal causes and conditions, like everything else in nature.

As noted previously, all usual kinds of experience, both sensory and introspective, are structured by memories, language, beliefs, and expectations, which cause us to assimilate even novel experiences, whether we want to or not. One of the names for the meditative practice I am describing here is "settling the mind in its natural state," which implies a radical deconstruction of the ways we habitually classify, evaluate, and interpret experience. The Buddhist hypothesis in this regard is that it is possible to so profoundly settle the mind that virtually all thoughts and other mental constructs eventually become dormant. The result is not a trancelike, vegetative, or comatose state. On the contrary, it is a luminous, discerningly intelligent awareness in which the physical senses are withdrawn and the normal activities of the mind have subsided.[17]

The culmination of this meditative process is the experience of the *substrate consciousness* (*ālaya-vijñāna*), which is characterized by three essential traits: bliss, luminosity, and nonconceptuality. The quality of bliss does not arise in response to any sensory stimulus, for the physical senses are dormant, as if one were deep asleep. Nor does it arise in dependence upon a pleasant thought or mental image, for such mental features have become subdued. Rather, it appears to be an innate quality of the mind when settled in its natural state, beyond the disturbing influences of conscious and unconscious mental activity.[18] A person who has achieved this state of attentional balance can remain effortlessly in it for at least four hours, with physical senses fully withdrawn and mental awareness highly stable and alert.

The quality of luminosity is not any kind of interior light similar to what we see with the eyes. Rather, it is an intense vigilance that has the capacity to illuminate, or make consciously manifest, anything that may arise within the space of the mind. To get some idea of what this is like, imagine being wide awake as you are immersed in a perfect sensory deprivation tank so that you have no experience of any of the five senses, or even of your own body. Then imagine that all your thought processes involving memory and imagination are put on hold, so that you are vigilantly aware of nothing but your own experience of being conscious. This is also analogous to "lucid dreamless sleep," in which one is keenly aware of being deep asleep, in a kind of wakeful vacuum state of consciousness.[19]

The empty space of the mind of which one is aware, once the mind has been settled in its natural state, is called the *substrate* (*ālaya*).[20] Due to the relatively nonconceptual nature of this state of consciousness, there is no distinct experience of a division between subject and object, self and other. Relatively speaking, the subjective substrate consciousness is nondually aware of the objective substrate, an experiential vacuum into which all mental contents have temporarily subsided. The mind may now be likened to a luminously transparent snow globe in which all the normally agitated particles of mental activities have come to rest. To draw an analogy from classical physics, virtually all the kinetic energy of the human psyche has been turned into potential energy, stored in this nondual experience of the substrate.

This natural, or relatively unstructured, state is permeated with an extraordinary amount of "creative energy" that has the capacity to generate alternative realities, such as whole dreamscapes that emerge from a state of deep sleep. To draw another analogy from contemporary physics, the substrate may be likened to the zero-point field, a background sea of luminosity permeated by an enormous amount of energy. This is the lowest possible energy state of the mind that can be achieved through such straightforward calming practices, and the energy of all kinds of mental activity is over and above that zero-point state.

For the normal mind, enmeshed in a myriad of thoughts and emotions, this zero-point field—substrate—of consciousness is unobservable, for we see things by way of contrast. Our attention is normally drawn to appearances that arise to the physical senses and mental perception, and they alone are real for us. But all such appearances originate from this zero-point field, which permeates all our experience. We are effectively blind to it, while the world of appearance arises over and above it. When sensory and mental appearances naturally cease, as in deep sleep, the mind is normally so dull that we are incapable of ascertaining the substrate consciousness that manifests.

The experience of the substrate is imbued with a relative degree of symmetry, and in this vacuum state reality does not appear in a structured form, either as a human psyche or as matter. This unstable equilibrium is perturbed by the activation of the conceptual mind, which creates the bifurcations of subject and object, mind and matter, which may be regarded as *broken symmetries*. When the fundamental symmetry of the substrate manifests in dreamless sleep, it is generally unobservable, and can only be retrospectively inferred on the basis of the broken symmetries of waking experience. But as mentioned before, as a result of continuous training in developing increasing stages of mental and physical relaxation, together with attentional stability and vividness, it is said that one may directly vividly ascertain this relative ground state of consciousness and observe how mental and sensory phenomena emerge from it in dependence upon a wide range of psychological and physical influences.

The mind gradually settles into the substrate consciousness as mental activities gradually subside, without suppression, throughout the course of this training. And in this process, memories, fantasies, and emotions of all kinds come to the surface of awareness. Our usual experience of our mental states is heavily edited and processed by the habitual structuring of the mind, so we tend to experience them in a way we regard as "normal." But in this training, the light of consciousness, like a probe into deep space, illuminates bizarre mental phenomena that seem utterly alien to one's past experience and sense of personal identity. As an analogy from contemporary astronomy, recall the million-second-long exposure of the Hubble Ultra Deep Field. Astronomers discovered in that region of deep space a zoo of oddball galaxies, in contrast to the classic images of spiral and elliptical galaxies. Some look like toothpicks, others like links on a bracelet, and a few of them appear to be interacting. These bizarre galaxies chronicle a period when the universe was more chaotic, when order and structure were just beginning to emerge.

Likewise, consciously exposing the deep space of the mind to thousands of hours of observation reveals normally hidden dimensions that are more chaotic, where the order and structure of the human psyche are just beginning to emerge. Strata upon strata of mental phenomena previously concealed within the subconscious are made manifest, until finally the mind comes to rest in its natural state, from which both conscious and normally subconscious events arise. This is an exercise in true depth psychology, in which one observes deep core samples of the subconscious mind, penetrating many layers of accumulated conceptual structuring.

Just as scientists expect that observations of the Hubble Ultra Deep Field will offer new insights into the birth and evolution of galaxies, so do Tibetan contemplatives believe that the experience of the substrate consciousness offers insights into the birth and evolution of the human psyche. Drawing on an analogy from modern biology, this may be portrayed as a kind of "stem consciousness." Much as a stem cell differentiates itself in relation to specific biochemical environments, such as a brain or a liver, the substrate consciousness becomes differentiated with respect to specific living organisms. This is the earliest state of consciousness of a human embryo, and it gradually takes on the distinctive characteristics of a specific human psyche as it is conditioned and structured by a wide range of physiological and, later, cultural influences. The substrate consciousness is not inherently human, for this is also the ground state of consciousness of all other sentient animals. Contrary to the hypothesis that consciousness ultimately emerges from complex configurations of neuronal activity, according to the Great Perfection (Dzogchen) tradition of Tibetan Buddhism, the human mind emerges from the unitary experience of the zero-point field of the substrate, which is prior to and more fundamental than the human, conceptual duality of mind and matter.[21] This luminous space is undifferentiated in terms of any distinct sense of subject and object. So this hypothesis rejects both Cartesian dualism and materialistic monism, and it may be put to the test of experience, regardless of one's ideological commitments and theoretical assumptions.

While resting in the substrate consciousness, one may deliberately direct attention to the past, gradually exercising memory until one can vividly and accurately recall events. Some Buddhists claim that within the distilled, luminous space of deep concentration, one may direct the attention back in time even before conception in this life and recall events in the distant past.[22] As far-fetched as this hypothesis may seem, it can be tested with carefully controlled experiments, assuming that the subjects involved are highly expert in this practice. By such rigorous examination, it should be a fairly straightforward process to determine whether such adepts' "memories" are accurate recollections from the past or mere fantasies.

Open-minded skepticism toward these claims—specifically, the kind of skepticism that inspires testing hypotheses in the most rigorous way possible—is healthy and appropriate for the scientific community. To the great detriment of science, however, the ideal of skepticism in the twentieth century has often degenerated into a kind of complacent closed-mindedness about any theory or method of inquiry that deviates from current mainstream science. Richard Feynman reminded us of the true ideal of scientific skepticism when he encouraged experimenters to search most diligently in precisely those areas where it seems most likely they can prove their own theories wrong.[23] Heraclitus, the sixth-century B.C.E. Greek philosopher known for his belief that the nature of everything is change itself, encouraged this open-minded attentiveness to novelty: "If you do not expect the unexpected, you will not find it, since it is trackless and unexplored."[24]

# 5

# A SPECIAL THEORY OF ONTOLOGICAL RELATIVITY

## The Illusory World of Perception

Philosophers and scientists have long recognized the illusory nature of perceptual appearances. When we observe the world around us, we see images, such as shapes and colors, that lack physical attributes. The visual image of the color red, for instance, doesn't have any mass or atomic structure. It isn't located in the external world, for it arises partly in dependence upon our visual sense faculty, including the eye, the optic nerve, and the visual cortex. There are clearly brain functions that contribute to the generation of red images, but no evidence that those neural correlates of perception are actually *identical* to those images. So there is no compelling reason to believe that the images are located inside our heads. Since visual images, or qualia, are not located either outside or inside our heads, they don't seem to have any spatial location at all. The same is true of all other kinds of sensory qualia, including sounds, smells, tastes, and tactile sensations.

To give another example, when we gaze at the stars, the white, sparkling points of light we see appear to exist far out in space. But the eyes can't project images into space, and those images don't come to us through space. The photons emitted by stars may travel thousands of years before striking our retinas, so before we experience visual images of the stars, they themselves have moved far from the points where we perceive them. So there are no stars or galaxies located in those regions.

We also experience qualia introspectively, without direct dependence upon any of our five physical senses. We can deliberately imagine shapes, colors, smells, and so on, and memories of such impressions also arise spontaneously. When we direct our attention "inward," we can observe discursive thoughts arising and passing away, and while asleep we can perceive dreamscapes and experience emotions, desires, and other mental processes much as we do during waking hours. These mental qualia arise in dependence upon specific brain functions, but, once again, there is no evidence that they are identical to their correlated neural events. While some mental phenomena falsely appear to exist inside our heads, others deceptively seem to exist in the outside world. For example, for centuries people have been "seeing" patterns among the constellations and imagining images among cloud formations. But these patterns and images don't actually exist where they appear, nor do they exist inside our heads. They have no mass or spatial location. In fact, they don't have any physical characteristics at all, for they are not constituted by the properties, relations, actions, or interactions of particles or fields.

In short, everything we observe extrospectively and introspectively consists of qualia, or appearances, and they are illusory in the sense that they seem to exist either in the external world or inside our heads, whereas in reality there is no compelling evidence that they are located anywhere in physical space. Although neuroscientists have identified many brain processes that have a causal role in generating subjective experience, there is no empirical evidence that any neural process is equivalent to any sensory or mental experience. It is often said that such qualia really consist of information that is processed in the brain, much as information is stored and processed in a computer. But the information that we think is inside a brain or a computer actually exists in the "eye of the beholder," which has no location in physical space. Information is not intrinsic to any computational system. The electrical state transitions of a computer are symbol manipulations that exist only relative to a symbolic interpretation by some designer, programmer, or user.[1]

All our immediate experience of the outer world and our minds consists of perceptual representations, none of which has physical qualities, and we commonly assume that those appearances correspond to real, independent objects in physical space. But this assumption is also deceptive. If our sense data resemble objects and thus represent them in the way that a movie of a scene represents the actual scene, then those qualia must closely resemble the physical objects to which they correspond. But sensory appearances, although dependent on physical processes, have no physical attributes themselves, whereas the things and events that make up the objective world have only physical attributes. So the two sets of phenomena can hardly be said to resemble each other. Moreover, sensory appearances are perceptible, but objectively independent physical objects are invisible to our senses. Those who believe in the "correspondence theory" of appearances and physical reality are saying there are two worlds that closely resemble each other, but the fact is that they have no qualities in common, so there is really no resemblance at all.[2]

Why then does anyone believe that sensory and mental qualia, including information itself, exist as physical objects in space? In modern society, in which so many people assume that the real world is the same as the physical world, this attitude is bound to influence the way we apprehend things. Psychologists are well aware of the fact that perception is largely a function of expectation, and if we believe that the objects of our perception are physical, then we expect them to be located in physical space.[3] The source of the illusion is our deeply ingrained tendency to reify the objects of perception, imagining them to be independent of our awareness of them.

## Ideas of Matter

The standard account of the history of experimental psychology places its origins in 1875, and most current, naturalist theories of the mind-body problem are largely based on the physics of that era, when material entities were defined as being located in space, causally connected to changes in their spatial environment, and endowed with mass. This idea goes back to Democritus, who declared that the objective world consists of atoms moving in space. Pythagoras and his followers proposed an alternative view, maintaining that all things are numbers, which they identified with geometrical forms. Plato built on this notion by proposing that the world of appearances emerges from an underlying realm of pure ideas. When Plato debated with materialists of his time, he showed that they didn't really know what they meant by "matter," then presented his own explanation of matter as consisting of immaterial structures.[4] In his view, each of the four elements—earth, water, fire, and air—exists in an ideal pure form, in a subtler dimension that transcends the world of the physical senses. The physical objects that make up our world of experience are impure, or mixed, forms of these four ideal elements. Solid objects represent coarse manifestations of the earth element; fluids represent the water element; manifestations of heat are impure expressions of the fire element, and gases correspond to the air element. Upon learning that only five perfectly symmetrical forms can be made from simple polygons (the triangle, square, and pentagon), Plato formulated a "theory of everything" in his work *Timaeus*, in which he added a fifth element, quintessence, of which space itself is made. According to this theory, all sensory qualia are indeed representations, not of physical entities but of ideal, immaterial forms of these five elements. Although the objects of perception appear to be physically real, this is an illusion, for their underlying reality consists of pure forms visible only to a higher, more refined type of perception.

Although physics through the nineteenth century corroborated the view of Democritus, with the twentieth-century revolution in the field, Werner Heisenberg concluded, "*modern physics takes a definite stand against the materialism of Democritus and for Plato and the Pythagoreans.*"[5] The primary reason is that the laws of nature mathematically formulated in quantum theory no longer describe elementary particles themselves but rather physicists' knowledge of quantum events. Quantum theory is not about objective physical reality, but about measurements of elementary particles.[6]

The world of physics was originally inspired by pure mathematics—a kind of immaculate conception, as it were—and during the era of classical physics, seemed to be made of up purely objective chunks of matter and fields. But over the past hundred years, this real, objective world has withdrawn back into mathematics, which is neither purely objective nor a purely subjective artifact of human imagination.

The laws of nature and their outcomes are expressed in the language of mathematics, and even the very structure of the universe is determined by unchanging qualities that can be encoded in a list of numbers called the "constants of nature." These include things such as the masses of the smallest subatomic particles, the strengths of the forces of nature, and the speed of light in a vacuum. The fabric of the universe and the pivotal structure of universal laws are seen to emerge from standards and invariants that transcend human experience.[7]

It is both astonishing and mysterious that mathematical theories can provide such accurate descriptions of the universe. To many physicists, this strongly implies the existence of a dimension of reality that transcends appearances of the physical world, and the ultimate simplicity of this mathematical reality enables scientists to investigate the world and have faith that their resulting understanding can converge on the truth.

## Psychophysical Coemergence

One of the greatest collaborations between a leading physicist and a leading psychologist took place in the twentieth century between Wolfgang Pauli and Carl Jung.[8] In their discussions and correspondence, Pauli often suggested that the mental and material domains might be epistemologically distinct, originating from an integral domain prior to the distinction of mind and matter. Jung called this more fundamental dimension of reality the *unus mundus*, from which archetypes can manifest as configurations of mental and physical phenomena.[9] He founded this idea on the assumption that the perceptual world, with all its categories of mind and matter, emerges from an underlying unity that transcends the physical senses. And he believed that the existence of this archetypal realm was essential to explain the causal connections between the psyche and the body.[10]

Pauli was equally intent on developing a new vision of an underlying reality that is inevitably symbolic, consisting of a fusion of human subjectivity and an objective order in the cosmos of which humans are only a part.[11] Like Jung, he conceived of symbols as *archetypal ideas*, which do not refer to explicitly accessible elements of everyday reality. Pauli proposed that mind and matter emerge by a breakdown of the psychophysical symmetry of the *unus mundus*. In this model, mental processes are psychic manifestations of archetypes and the physical laws are physical manifestations of archetypes, and he speculated that there should be natural laws, with an inner correspondence, governing both emergent domains.

While such a hypothesis appears implausible in terms of classical physics, since twentieth-century physics considered matter an abstract, invisible reality, he felt that such a psychophysical monism had become more feasible. Nevertheless, like Copernicus, Pauli recognized that his ideas ran against the dominant ideology of his contemporaries, so he did not express them out of fear of ridicule from his scientific peers.

The fundamental idea here is not that mental phenomena emerge from complex configurations of matter, as is widely assumed today, but rather that the distinction of mind and matter emerges from an underlying reality of archetypes. This concept was not unknown during the scientific revolution. Benedict de Spinoza, for example, proposed that there is one fundamental substance, a *causa sui*, from which all particular manifestations of mind and matter derive. Variations on this theme have also been suggested by a growing number of twentieth-century physicists. David Bohm is well known for his theory of the implicate order, existing prior to the distinction of mind and matter, which is on the level of an explicate order. In our perceptual world, there seem to be a "mental pole" and a "physical pole," but the deeper reality is something beyond either.[12] Physicists Eugene Wigner and Bernard d'Espagnat have also advocated similar views of an underlying reality.[13]

## A Holographic Universe

Recently some physicists have proposed that our universe, which we perceive to have three spatial dimensions, may actually be emerging from a two-dimensional surface, like a hologram, suggesting that our everyday perception of the world as three-dimensional is either an illusion or merely one of two alternative ways of viewing reality.[14] One of the bases for this theory has to do with the mysterious properties of black holes, which imply how much information a region of space or a quantity of matter and energy can hold. This notion, known as the "holographic principle," was first proposed in 1993 by Nobel laureate Gerard 't Hooft and later developed by Leonard Susskind, known for his discovery of string theory. They contend that our illusory, three-dimensional world is completely described by a physical theory defined only in terms of a two-dimensional "boundary" of our universe. Or, if we consider our world to consist of four dimensions, having three-dimensional volume and extending into the fourth dimension of time, they speculate that there is an alternative set of physical laws, operating on a three-dimensional boundary of space-time somewhere, that would be equivalent to our known four-dimensional physics.

Other physicists have suggested variations on this holographic principle, which has not yet achieved the status of a physical law, and many believe that such a theory in its mature form will be concerned not with fields or even with space-time, but rather with information exchange among physical processes. This would imply that the physical world essentially consists of information that becomes embodied in configurations of mass-energy and space-time.[15]

Physicist George Ellis has proposed a fourfold model of reality, consisting of matter and forces, consciousness, physical and biological possibilities, and mathematical reality.[16] All of these levels of existence are ontologically real and distinct, but are related through causal links. Language and symbols exist as nonmaterial effective entities, created and maintained through social interaction and teaching. They are not contained in any individual brain, nor are they equivalent to brain states, though they may become embodied in neural circuitry and other complex systems, such as molecular biology, language and symbolic systems, individual human behavior, social and economic systems, digital computer systems, and the biosphere. In all these systems, vast quantities of stored data and hierarchically organized structures process information in a purposeful manner, particularly through implementation of goal-seeking feedback loops. This produces emergent behavior, in which the behavior of the whole is greater than the sum of its parts, and cannot even be described in terms of the language that applies to the parts.

Like Pauli, Ellis advocates the existence of a Platonic world of abstract realities that can be discovered by human investigation but are independent of human existence. Such realities are not embodied in physical form but can have causal effects in the physical world. Major parts of mathematics, such as rational numbers, zero, and irrational numbers, are discovered rather than invented and therefore have an existence of their own. He writes, "They have an abstract character, and the same abstract quantity can be represented and embodied in many symbolic and physical ways. They are not determined by physical experiment and are independent of the existence and culture of human beings."[17] This Platonic world, he believes, in some way underlies the world of physics and has the power to control the behavior of physical phenomena.

This belief stems from an awareness of the "unreasonable power of mathematics" to describe the nature of physical processes. Apart from the hypothesis that the world is constructed on a mathematical basis, it is hard to explain why the behavior of matter can be accurately described by equations of the kind encountered in present-day mathematical physics. And it is equally hard to fathom why all matter has the same properties throughout the known universe. As discussed in the first chapter of this volume, the standard naturalist accounts of the universe have no explanation for the mathematical nature of physical processes; mathematical theories alone do not explain the emergence of a physical universe; physical theories alone do not explain the emergence of life in the universe; and biological theories alone do not explain the emergence of consciousness in living organisms. Each of these branches of science provides partial and incomplete explanations, and there are always multiple levels of explanation that all hold at the same time. This implies that no single explanation is complete, so one can have a top-down system explanation as well as a bottom-up explanation, both simultaneously applicable.[18] While so-called higher-level explanations rely on the existence of the lower-level explanations, they are of a different nature than, and not reducible to, the lower-level ones. In this sense, the higher-level explanations are deeper. Like Pauli, Susskind, and other physicists cited in this chapter, Ellis speculates that information may be the key to understanding the origins of specific laws of physics and the specific initial conditions of the universe.

Many other theoretical physicists and mathematicians are drawn to the idea that information about abstract objects is acquired by means of a faculty of mathematical intuition. Roger Penrose, for example, agrees with Ellis that mathematical realities are not determined by physical experiment but arrived at by mathematical investigation. They have an abstract rather than embodied character, and can be represented and embodied in many symbolic and physical ways. They are independent of the existence and culture of human beings, for mathematicians believe that their features would be discovered by intelligent beings anywhere in the universe if their mathematical understanding were sufficiently advanced. Humans are one species that has discovered this dimension of existence, which we represent in our mathematical theories. Those representations are cultural constructs, but the underlying mathematical features they represent are not, for they are truly discovered as are physical laws.

Penrose believes exceptionally gifted mathematicians and theoretical physicists are able to "visit" a Platonic world of pure ideas, where they make genuine discoveries.[19] Indeed, not only mathematical understanding but also human musical, artistic, and aesthetic creativity and appreciation, he suggests, come from contact with this realm. In line with other physicists mentioned above, Penrose has developed a universal mathematical framework simultaneously representing the materialistic world of physical reality and the Platonic world of mathematical reality.

With respect to all formulations of a Platonic realm, or any fundamental dimension of existence from which physical and mental phenomena coemerge, the challenge is to understand how such a realm pertains to contemporary physics and how it interacts with the brain. All such hypotheses may be classified under the general category of a special theory of ontological relativity, in which our familiar world of mental and physical phenomena exists only relative to this underlying, unitive domain. For such a theory to be deemed scientific, it must somehow be testable through experience, and this will require an expansion of our current understanding of both physics and the mind.

# 6

# HIGH-ENERGY EXPERIMENTS
# IN CONSCIOUSNESS

## Parallels in Particle Physics

Throughout the nineteenth century, scientific speculations about the existence and nature of atoms were largely metaphysical, with physicists and chemists philosophically arguing their different views. The first compelling proof of the existence of atoms appeared in 1908, when Jean Perrin compared the effects of gravity and Brownian motion (random movement of microscopic particles suspended in liquid) on minerals dissolved in water, and was thereby able to infer the mass of the surrounding molecules causing this motion. Three years later, Ernest Rutherford directed a stream of alpha particles (later identified as positively charged helium atoms) into thin sheets of gold foil. The particles were so minute that almost all of them should have passed unimpeded through the foil, but Rutherford found that a significant number were deflected at various angles. From this he inferred that there was a hard core in a gold atom, which he called its nucleus.

In the twenty-first century, breakthroughs in particle physics will likely come from experiments conducted with the Large Hadron Collider, a particle accelerator currently being constructed at CERN (European Organization for Nuclear Research) and scheduled to start operation in late 2007. This will become the world's largest particle accelerator, consisting of a tunnel 27 kilometers in circumference. With it, physicists hope to answer the following basic questions:[1]

- What is mass?
- What is the origin of the mass of particles?
- Why do elementary particles have different masses?

- What are dark matter and dark energy, which are believed to make up 95 percent of the universe's mass?
- Do superparticles, particles related to more standard particles by supersymmetry, exist?
- Are there extra dimensions, as predicted by various models inspired by string theory, and can they be observed?
- Are there additional violations of the symmetry between matter and antimatter, two types of matter that become annihilated when they come into contact with each other?

High-energy physics seeks to discover basic principles that underlie the workings of the physical universe by exploring the building blocks of matter and forces among them, but profound questions remain for which scientists hope the new technology will provide answers. It is remarkable, in retrospect, that after 400 years of progress in physics, we still don't know what matter or energy is. Skeptics may be dubious about the likelihood of particle colliders shedding light on the nature of the observer, which experiments have revealed plays an essential role in all quantum phenomena, from the interactions of elementary particles to the formation of the galaxies. It seems even less likely that such technology will reveal the nature of a Platonic realm of archetypes, if it exists.

To scientifically investigate the nature and origins of mass-energy, the role of the observer, and the possible existence of a dimension of reality that precedes and transcends mind-matter distinctions, new lines of inquiry may be needed. In light of the past successes and limitations of mainstream physics, current research protocols appear inadequate for studying the interface between the mind of the observer and physical phenomena, including those occurring in the brain. For a true breakthrough in such research, I believe that we must explicitly include rigorous mental training and the transformation of consciousness, and this is where the contemplative traditions of the world may have much to offer.

Skeptics may quite rightly counter that this approach would merge physics with metaphysics. But a mingling of empirical science with philosophical speculation characterizes the entire history of physics. Before the twentieth century, the origins of the universe, the existence of other galaxies, and the nature of atoms and elementary particles were matters of metaphysical speculation. Only with the development of appropriate technologies were these areas of inquiry moved from the domain of metaphysics to science. While tremendous technological advances have been made in observing physical phenomena that existed billions of years ago, billions of light-years away, and in the inner core of the atomic nucleus, no comparable advances have been made in probing the origins of consciousness, observing the depths of the space of consciousness, or examining the human mind.

## Requisites for High-Energy Consciousness Research

To engage in high-level consciousness research, one must first develop certain mental qualities and a way of life conducive to such inquiry. The normal, untrained mind is addicted to sensory and intellectual stimulation and is consequently prone to emotional vacillations, which seriously disrupt the advanced research outlined in this chapter. To succeed in experimenting with one's own consciousness, it is first necessary to cultivate confidence, effort, mindfulness, understanding, and concentration, and these qualities must be supported by a way of life that nurtures the cultivation of the mind, rather than undermining it. There is no way to separate lifestyle from meditative training, as can be done to a large degree with lifestyle and scientific research.

Trainees must have confidence in the practices, in the competence and altruism of their instructor, and in their own ability to engage in the training. However, too much confidence may lead to overexcitement, and this can impair judgment, which in turn interferes with a proper level of enthusiasm, mindfulness, and concentration. It is common for novices to try too hard, and it has been found that excessive effort agitates the mind and may result in physical problems as well. A common metaphor in the Buddhist tradition is to tune the attention as one would string a lute—not too tight and not too loose—for too much effort results in nervous imbalances, and too little leads to dullness and lethargy. In particular, confidence must be balanced with understanding, and effort must be balanced with concentration. Such mental balance is to be achieved by preventing agitation due to excess confidence, effort, or understanding, and laziness due to excess concentration. For this reason, mindfulness is necessary at all times, for it protects the mind and keeps the object of meditation from being lost.[2]

For all meditative research, a wholesome way of life, based on ethical discipline, is of paramount importance. Otherwise, in the course of the experiments described below, the mind is bound to succumb to remorse and agitation, which obstruct the development of meditative concentration.[3] The essence of ethical discipline is twofold: avoiding any behavior of body, speech, and mind that is injurious to oneself or others, and devoting oneself to conduct that serves the well-being of oneself and others. To give a slightly more elaborate account of the preliminary training required as a foundation for such meditative practice, traditional Buddhist sources cite four elements: ethical discipline, restraint of the sense faculties, mindfulness and introspection, and contentment.[4] Basic requisites also include a suitable diet, clothing, and, when needed, medication.

Much as research in physics, biology, and psychology requires a well-designed laboratory, sophisticated, meditative research into consciousness requires a conducive environment. According to traditional Buddhist sources, an appropriate facility, or consciousness research laboratory, for this training should have five qualities:[5] be easily accessible, so that trainees can acquire food, clothing, and medicine; be free from danger caused by humans and animals; be aesthetically agreeable and healthy; be inhabited by good companions who are ethically disciplined and like-minded; and be serene and quiet, with little commotion by people during the daytime and little noise at night.

To ensure the greatest possibility of success in this training, it is also indispensable to practice under the guidance of a qualified instructor. In this day and age, such teachers are hard to find, but at the very least a teacher should have an altruistic motivation and greater understanding and experience in such practice than the student.

## Empirical Research Into the Dimension of Archetypal Forms

The following series of experiments belongs to the disciplines of deep psychology and physics and opens the possibility of experientially exploring an implicit, symbolic dimension of reality proposed by the scientists cited in the preceding chapter. Buddhism also posits the existence of such a dimension of pure forms, which will be discussed later in this chapter. These experiments are not designed to validate a particular theory, but rather to provide experiential data to support the existence of any such domain of reality from which our familiar world of mind-matter distinctions emerges.

These experiments in consciousness focus on earth, water, fire, air, and space, closely paralleling the five elements described by Plato. The general strategy is to focus the mind initially on physical emblems of these five elements, which correspond to basic states of physical phenomena: solid, fluid, heat, movement, and space. Through those preliminary exercises, the mind is focused single-pointedly on a mental image that reflects the earlier sensory impression of the emblems. So far, this constitutes a straightforward training of the attention. But the remarkable discovery that is allegedly made after becoming thoroughly adept in these first two stages of practice is that archetypal forms of the elements eventually dawn in one's consciousness. These symbolic forms are not memories of earlier impressions of the five elements, subjective fantasies, products of our own subjective awareness, or objective, preexisting archetypes in a higher dimension of reality. Like so many phenomena "discovered" in modern physics, they arise in the interface between subjective and objective elements of reality.

Let us begin with an experiment involving the earth element.[6] First, we form a disc of homogenous clay, free of straw, pebbles, and other imperfections, about 10 inches in diameter, and lay it on a smooth surface in front of us. Seated comfortably on a slightly elevated platform about five feet away, we gaze intently at this symbolic representation, mentally viewing it as universally representing solids everywhere. We calmly observe this emblem as if viewing our own reflection in a mirror. We do not concern ourselves with its color, but attend solely to its quality of solidity, and conceptually bear this idea in mind, even mentally repeating "solid, solid." We continue looking at the clay disc until we are thoroughly familiar with it, then intermittently close our eyes and attend to the mental image corresponding to its visual appearance.

We do this repeatedly with strong concentration until this mental image appears as steadily and vividly as if we were still seeing the disc with our eyes. As soon as we reach this stage, we disengage from the physical emblem of solidity, move to a quiet place indoors, and continue focusing solely on the mental image. If at any time we have trouble recalling the image, we return briefly to its physical representation until the mental image is restored. As we continue this exercise, our mind will eventually settle into a profoundly stable, vivid state of focused attention. While our physical senses are completely withdrawn and our mind is single-pointedly focused on this image, eventually an archetypal symbol of solids, orders of magnitude subtler, will spontaneously break through.

Not everyone experiences this subtler form in the same way. We are not discovering a preexisting, objective symbol that exists independently of our mind, nor a mere figment of our imagination. It has an archetypal quality, yet its specific manifestation is related to our own perception. But unlike any visual image or mental representation of such an image, this rarified icon has no color or shape. Those who have not reached this state of focused attention cannot imagine what it is like, but those who have can converse about what they have experienced, much as highly trained mathematicians converse among themselves in a semiprivate language.

The realization of the substrate consciousness, discussed in chapter 4, is a kind of portal that provides access to this subtle, archetypal dimension of existence. One's ordinary psyche needs to be shut down, in a dormant state, before one can cross the threshold into this realm that underlies the dualistic world of mind and matter. The archetype of solids first appears only fleetingly, then disappears as the awareness slips back into the substrate consciousness, devoid of content. But if one steadfastly fixes the attention on the icon, this deep state of concentration gradually stabilizes until one can remain in it for up to twenty-four hours without a break, physical senses utterly withdrawn and mind unencumbered by any perturbing influences.[7]

The dimension from which this iconic representation of solids emerges is that of purely archetypal forms, from which the physical and mental universe as we know it emerges. Materialists would assume that the icon emerges from the brain, and philosophical constructivists would assume that it is a result of various influences in one's genetic and social background. But Buddhists hypothesize that it emerges from an interaction between the individual psyche and this higher dimension of pure forms.

Between meditation sessions, it is said that we retain an exceptional degree of mental and physical pliancy and fitness, which causes us to be naturally inclined to act in ways conducive to our own and others' well-being. Adepts report that due to a radical transformation in their nervous systems, they experience a kind of bodily fitness such that they have no feelings of physical heaviness or discomfort and are saturated with a sense of bliss. Another alleged result of this practice is an unprecedented degree of mental fitness, so that one is fully in control of the mind, virtually free of sadness and grief, and continuously experiencing a state of well-being.

The genuine happiness that emerges as a trait effect of having settled the mind in its natural state is fundamentally unlike "hedonic" pleasure that arises in response to chemical, sensory, aesthetic, and intellectual stimuli. This unprecedented sense of well-being arises because the mind has been brought to a deep equilibrium, in which our "psychological immune system" has been enhanced so that we rarely succumb to the mental disturbances of craving, hostility, anxiety, or depression. Such happiness is directly related to insight into the nature of our own mind, which is directly relevant to self-knowledge; as explained above, it arises only within the context of a virtuous way of life. This deepening of happiness and understanding reciprocally influences the quality of life as a whole, inhibiting unwholesome and unethical behavior while supporting virtues of all kinds.[8]

Once we have settled our mind in its natural state, we may initially gain experiential access to the realm of pure forms by focusing on the earth element, or we may start with any of the other elements of water, fire, air, or space.[9] Traditional Buddhist sources cite other emblems that may be used to access this dimension of existence, and they also recommend specific emblems most suitable for people with different kinds of temperament.[10] The initial object for focusing on the water element may be a bowl, bucket, or well of pure, clear water. As in the preceding experiment, we concentrate on the concept of fluid and mentally recite, "fluid, fluid," until a mental image corresponding to the physical emblem arises and we can sustain it at will. Eventually, the archetypal form of fluid will arise as in the case of the earth element.

For the fire element, we use a candle, or any other flame we remember seeing. One strategy is to make a screen with a circular hole in it about one foot across. We put the screen in front of a wood or grass fire, so we see only the flames through the hole. Ignoring the smoke and the burning fuel, we concentrate on the concept of fire until the mental image arises, and then develop it in the usual way. We may focus on the air element by way of our sense of touch or sight. For example, we may concentrate on the sight of leaves or branches moving in the wind, the sensation of a breeze touching our body, or the passage of our breath at the apertures of our nostrils. In any of those exercises, we concentrate on the concept of air until its mental image is stabilized, then proceed as before. Finally, to access the archetype of space, we first direct our attention to the space in a doorway, window, or keyhole. Alternatively, we may make a circular hole in a piece of board, about 10 inches in diameter, hold the board up so we see only the sky through the hole, no trees or other objects, and concentrate on the space within that circle. In the meantime, we absorb our attention in the concept of space and continue as before.[11] According to traditional Buddhist sources, each of the above methods provides experiential access to emblematic representations, or archetypes, of the whole quality of the elements they symbolize.[12] If one has already completed the training in settling the mind in its natural state, this will be relatively easy to accomplish and will not take long. Without prior training in developing one's attention skills, it may take 5,000 to 15,000 hours to complete these experiments pertaining to the archetypal correlates of the five elements.

The above method for experientially exploring the archetypal realm of pure ideas seems to have been quite prevalent in India and Southeast Asia for the first millennium after the Buddha, but over the past 1,500 years, such practice has declined, especially following the European domination of Asia. Fortunately, such training is still done in a few establishments in Asia, such as the Pa-Auk Tawya Meditation Center in Myanmar.[13] Hopefully, experiments in high-energy consciousness will soon be conducted in the East and the West, coupling these traditional practices with rigorous scientific protocols.

The primary reason Buddhists have traditionally engaged in such practices is to alleviate five afflictive mental traits that perturb the equilibrium of the mind, resulting in a wide array of mental imbalances. These "five hindrances" are sloth and torpor, doubt, ill will, distraction and agitation, and sensual craving. Each is gradually overcome through the cultivation of five corresponding qualities of meditative stabilization: initial mental application, sustained mental application, joy, happiness, and concentration. To summarize this process:[14]

- The factor of initial mental application counters the combined hindrances of sloth and torpor, and it applies the mind to the object of concentration.
- The factor of sustained mental application counters the hindrance of doubt and keeps the mind continually engaged in the exercise.
- The factor of joy counters the hindrance of ill will and increases interest in the object. Through the course of this training, joy gradually increases in five stages, manifesting as:[15]
    slight sense of interest;
    growing interest, which is momentarily keener;
    absorbing interest;
    thrilling sense of interest;
    intense joy, which saturates one's whole mind and body and is associated with a powerful state of concentration.
- The factor of happiness counters the combined hindrances of distraction and agitation and helps to concentrate the mind.
- The factor of concentration counters sensual craving and arises in dependence upon the above four factors of stabilization.

When the archetypal form of any of the elements first arises, the five hindrances are temporarily suppressed, but they may still dominate the mind on occasion in between meditation sessions. However, if we persist in learning to stabilize these symbolic forms, the mental imbalances become largely dormant for as long as we maintain that degree of meditative concentration. With the body saturated with well-being and the mind settled in a state of equilibrium, Buddhist contemplatives claim that one's well prepared to begin exploring the nature of reality in order to completely and irreversibly liberate the mind from all its afflictive tendencies.[16]

Experiential access to the realm of pure forms is an important element of such inquiry. According to Buddhism, this dimension of existence is not simply a subjective state of consciousness, but exists independently of the human mind. In fact, traditional Buddhist sources provide separate accounts of the form realm (*rūpa-dhātu*): ontological descriptions of it as a preexisting dimension of the natural world[17] and epistemological accounts of how to ascertain it through meditative training.[18] According to some schools of Buddhism, this archetypal realm consists of four subdomains pertaining to the four elements of earth, water, fire, and air. Successive implicate orders are postulated, with the physical world as we experience it emerging from the form realm and the form realm emerging from a subtler formless realm (*arūpya-dhātu*).[19] Such descriptions bear some similarities to the Pythagorean hypothesis of a realm of mathematical archetypes, which Plato expanded to include both mathematics and qualitative pure ideas. It would be fascinating to explore how closely the form realm as described by Buddhists corresponds to the Platonic realm of pure ideas, and whether the Buddhist formless realm might somehow correspond to the Pythagorean dimension of pure mathematics. Without experiential research, all such comparisons are only speculative.

Over the past 2,500 years, generations of Buddhist scholars and contemplatives throughout Asia have claimed that it is possible to develop paranormal abilities of mind over matter and extrasensory perception after achieving mastery of these archetypes of the elements.[20] While early Buddhists expressed ambivalence, if not downright disdain, regarding the use of such abilities (which are nevertheless cited in ancient Buddhist sources),[21] the later Mahāyāna and Vajrayāna schools of Buddhism endorsed them as long as they were used with wisdom and a compassionate motivation.[22] All of the most renowned scholars and contemplatives of these Buddhist traditions seem to have taken for granted that such abilities are real. The question of whether this is simply an extremely persistent superstition or a factual conclusion based on centuries of empirical evidence awaits scientific study.

## Scientific Evaluation

"Platonic physics" is based on Plato's admonition that the mathematical forms of experience are somehow more real than the physical world of our everyday experience and of scientific inquiry, but until now the evaluation of such ideas has often fallen short of established criteria of scientific rigor.[23] If the above theories and experiments are ever to gain scientific credibility, they must be examined with the greatest care and precision. In science, the true hallmark of the "real" is the observable consequences that a community of experienced investigators agrees occur in actual practice, and this is precisely the claim made by generations of Buddhist contemplatives. The primary criterion of good science is that a theory has been repeatedly tested by measurements—no matter how difficult the testing may prove—and found to be in excellent accord with predicted results.

But any scientific exploration of reality that includes subjective experience is bound to violate the "taboo of subjectivity," namely the insistence that any scientific theory must refer to purely objective phenomena that exist independently of our minds.[24] Scientists must indeed do all they can to avoid the influence of subjective biases in their research, such as favored theories or unexamined assumptions. But the taboo of subjectivity is exactly such a prejudice, traceable to the metaphysical dualism of Descartes. This view accords well with the classical physics of the nineteenth century, but as we have seen, it is incompatible with quantum theory and further advances in twenty-first-century physics.

Researchers must be as skeptical of their own uncorroborated assumptions as they are of novel theories and modes of empirical inquiry. Such skepticism plays a role like that of death in the evolution of science—only the strongest theories survive and propagate with new generations of scientists. Having introduced this evolutionary metaphor, physicist Michael Riordan acknowledges that speculative theorizing plays a crucial role by helping to ensure that science investigates the many philosophical niches where truth might lurk. But he adds that hypotheses resulting from such wide-ranging explorations of possible theory space must ultimately lead to testable consequences—a process that may take years, even decades—if science is to advance. Otherwise, theorists are doing metaphysics, not physics.

In general, scientific research follows a four-step process: study the relevant phenomenon, formulate a new theory, use the theory to predict observations that we should be able to make if the theory is correct, and look for these predicted observations. Within that context, Ernan McMullin has proposed four criteria of "complementary virtues" for evaluating a new theory:[25]

Internal virtues: logical consistency, coherence (or "naturalness," absence of ad hoc features), causal specificity

External virtues: consonance with other parts of science and (more controversially) consonance with broader worldviews (metaphysical principles of natural order, for instance)

Diachronic virtues: revealed over time as the theory develops and meets new challenges: fertility, consilience (unification of scientific domains previously thought disparate)

Uniqueness: the absence of credible theoretical alternatives

All the above criteria for evaluating a scientific theory have to do only with its claim to truth. But governments and industry are also concerned with the usefulness of theories and empirical research for advancing technology. These twofold purposes, which boil down to knowledge and power, go back to Francis Bacon's ideal of acquiring scientific knowledge of the physical world in order to control it for human ends.[26] While this approach has yielded innumerable benefits for humanity, it has also come with a high cost. As German physicist Carl Friedrich von Weizsäcker comments, there is a straight line from the physics of Bacon and Galileo to the atom bomb.[27] And in his 1946 lecture "The History of Nature," he argued that the scientific and technological world of modern times is the result of man's venturing knowledge without love.[28] This approach to scientific inquiry, largely devoid of ethics and altruism, has played a major role in the tremendous advances in science and technology made during the twentieth century, changing human society and our natural environment in countless ways. But this century has also produced the greatest inhumanity of man against man and the greatest degradation of the natural environment, including the extinction of countless species of plant and animal life, in human history. Prior to the advent of modern science, the human species adapted to environmental changes the old-fashioned way—through random genetic mutations. But now our natural environment has changed so rapidly, largely as a result of advances in science and technology, that there is no way humans can adapt through purely biological processes. If we are to survive our lopsided growth in knowledge and power, which has not been complemented by a comparable growth in ethics and social responsibility, then we must take our further evolution into our own hands. We must grow in wisdom and compassion or face the real possibility of extinction.

In the contemplative science set forth in this volume, a further set of "complementary virtues" may be proposed for evaluating any scientific theory. For assessing the truth of a theory, the earlier criteria proposed by McMullin may be adapted without modification. But in addition, two other criteria may be introduced. Any line of research—whether investigating elementary particles, mapping the human genome, or researching the depths of human consciousness—may be judged in terms of the extent to which it contributes, or is likely to contribute, to human flourishing, or genuine happiness. What is its potential value in terms of alleviating physical and mental illness, and how might it help develop exceptional degrees of physical, psychological, and spiritual well-being? The second additional criterion is the efficacy of scientific research for developing human virtues such as wisdom and compassion. Using these three criteria—truth, genuine happiness, and virtue—to evaluate theories and methods of inquiry promises to put a human face on the impersonal countenance of science. And it may contribute not only to our survival as a species but also to our conscious evolution in ways never before imagined.

# 7

# A GENERAL THEORY OF ONTOLOGICAL RELATIVITY

## Philosophical Precedents

The preceding two chapters have presented theories and experiments pertaining to our perceptual world of physical and mental phenomena as they are detected by the instruments of science, sensory experience, and introspective awareness of the mind. All such manifestations of mind and matter, I have proposed, emerge from and exist only relative to a subtle dimension of existence of pure forms, or archetypal symbols. I have called this a special theory of ontological relativity. In this and the next chapter I shall present theories and experiments concerning a general theory of ontological relativity that encompasses all possible phenomena, both perceptual and conceptual.

The essence of this view is that all phenomena can be posited to exist only in relation to a cognitive frame of reference. Antecedents of this hypothesis appear in the writings of Wittgenstein, who proposed that the truth of empirical propositions can be validated only with respect to a frame of reference; he likened this philosophical move to the step taken in Einstein's relativity theory.[1] The concept of ontological relativity was further developed by the American philosopher Willard Quine. Any theory, he maintained, makes sense only relative to a "background theory" and to its translation into that background theory. In this sense, ontology is doubly relative.[2]

In addressing the relativity of scientific theories, some philosophers have drawn a distinction between "observable" and "unobservable" entities, claiming that the former do not depend on a theoretical interpretation, whereas the latter do.[3] According to this view, when we make "ordinary observations," what we perceive is theory-independent and is a function of facts about us as organisms in the world.[4] The limitation of making such observations is that we have to start our investigations of nature somewhere, and this means that we have to rely on our previous understanding and our language.[5] In other words, all our observations, ordinary or otherwise, make sense only in relation to our background theory.

A serious problem for anyone making such a strong demarcation between "observable entities" and "theoretical entities" is deciding where to draw the line. As noted earlier, with years of theoretical and empirical training, scientists learn to observe things that no one else can apprehend. Psychologists have long known that ordinary perceptions are strongly influenced by memories and expectations, which enable us to recognize even novel objects and events within a familiar conceptual framework. Since no ordinary perceptions are free of such conceptual influences, the distinction between observable and unobservable entities appears arbitrary. To put it another way, that line itself is an unobservable, theoretical entity, and as such, it has no objective existence independent of our background assumptions.

According to the general theory of ontological relativity advocated here, the truth of a theory cannot be thought of in terms of a "correspondence" with some absolutely objective reality. The reason is that the objects posited in a theory do not exist independently of the procedures for making observations of the world and identifying stable, invariant elements in it. Physicists still do not know what either mass or energy is, as independently existing entities in the objective world. For all practical purposes, they simply define mass and energy in terms of the invariants that they extract from their methods of observation.[6]

In calling this a theory of ontological relativity, I do mean to suggest a parallel with Einstein's famous theory. One of his most profound discoveries was that there is no absolute inertial frame of reference—no absolute space or all-pervasive medium such as the ether—in relation to which the motions of physical bodies can be measured. Likewise, I propose that there is no theory or mode of observation—no infallible method of inquiry, scientific or otherwise—that provides an absolute frame of reference within which to test all other perceptions or ideas. One person's background theory may be someone else's foreground theory, and there is no universal, absolute code of translation with which to make sense of one theory in terms of another.

According to Einstein, the speed of light is invariant across all inertial frames of reference. Anyone anywhere traveling at any velocity always perceives light as traveling at the same speed, regardless of the direction it is traveling. Einstein's special theory of relativity pertains to inertial frames of reference traveling in straight lines at constant speeds, and in his general theory he expands this principle to include all frames of reference, whatever their speed or direction. Both theories are as much about invariants as they are about relativity. In the theory of ontological relativity, there is one truth that *is* invariant across all cognitive frames of reference: *everything that we apprehend, whether perceptually or conceptually, is devoid of its own inherent nature, or identity, independent of the means by which it is known.* Perceived objects, or observable entities, exist relative to the sensory faculties or systems of measurement by which they are detected—not independently in the objective world. This is the broad consensus among psychologists, neuroscientists, and physicists. For example, colors exist relative to the visual faculty that sees them, and sounds exist relative to the auditory faculty that hears them. Nevertheless, in our intersubjective experience, humans apprehend colors and sounds in similar ways, and we define them in terms of the invariants that we extract from our methods of observation. This allows for true statements to be made about such qualia that are independent of any specific subject. *But this doesn't mean that these phenomena are independent of all subjects or modes of perception.*

Likewise, so-called unobservable, theoretical entities, such as electromagnetic fields, dark matter, and the spin of elementary particles, exist in relation to the conceptual faculties and frameworks by which they are apprehended. Scientists in different laboratories using different research methods may agree on the inferred qualities of such entities, and this allows for true statements to be made about them that are independent of any specific scientist or laboratory. But this does not mean that such theoretical entities exist independently of any system of measurement and any background theory. Even if a great number of people look at a fire engine and see it as red, that doesn't mean the color exists independently of their visual faculties. Likewise, even if a great number of scientists detect the presence of a subatomic particle, interpreting it within the framework of a common background theory, that doesn't mean the particle exists independently of their theories and systems of measurement. The only invariant across all these cognitive frames of reference is that nothing exists by its own nature, independent of all means of detecting it or conceiving of it. In other words, there is no way to separate the universe we know from the information we have about it.

# A World of Information

Formal information theory originated in 1948 and measures information content in terms of entropy, which has long been a central concept of thermodynamics, the branch of physics that deals with heat. Thermodynamic entropy is commonly associated with the degree of disorder in a physical system. More precisely, such entropy is characterized in terms of the number of distinct microscopic states that the particles composing a chunk of matter can be in while still looking like the same macroscopic chunk of matter. According to current information theory, the entropy of a message is quantified as the number of binary digits, or bits, needed to encode it. This objective measure has been very useful in science and technology, but it says nothing about the value or meaning of information, which is highly dependent on context.

One application of this theory that has generated heated debate pertains to the conservation of information in the creation of black holes, regions of space-time with a gravitational field so strong that nothing can escape them—not even light.

According to Stephen Hawking, all information is irretrievably lost down the hole, implying that the entropy of a black hole is associated with lost information. But other physicists, including Gerard 't Hooft, argue that when a black hole disappears, it must eventually give back to the universe all the information it swallowed. This implies a universal conservation of information. In principle, 't Hooft maintains, it should be possible to retrieve all the black hole's information content by examining all the degrees of freedom on its two-dimensional surface. This model invokes the analogy of a hologram, in which a three-dimensional image is created by shining a laser on a two-dimensional plate. Applying this metaphor to the universe at large, 't Hooft characterizes physical phenomena not by the three-dimensional volume they occupy in space but by a "projection" of their degrees of freedom on a two-dimensional area. This idea, as mentioned previously, is called the holographic principle.

In his characteristically bold fashion, John Wheeler reappraises the scientific concept of information by proposing that any true observation of the physical world must not only produce an indelible record but also impart *meaningful information*. The act of measurement implies a transition from the realm of mindless stuff to the realm of knowledge, so a measurement must record a bit of information that means something. Information, in other words, should inform, and for this to take place there must be someone who is informed. So Wheeler expands the scientific notion of information beyond the mindless context of entropy to include the observer-participant. Information is no longer a purely objective entity but includes the subject as well.

In Wheeler's provocative vision, the universe consists of a "strange loop," in which physics gives rise to observers and observers give rise to physics. But he goes beyond this two-way interdependence and turns the conventional explanatory relationship *matter → information → observers* on its head, placing observership at the base of the explanatory chain *observers → information → matter*. This line of thinking led to his famous "it from bit" dictum—the "it" referring to a physical object such as an atom, and the "bit" being the information that relates to it.[7] According to this theory, the universe is fundamentally an information-processing system from which the appearance of matter emerges at a higher level of reality.

A growing number of leading physicists, including Carl Friedrich von Weizsäcker, have come to characterize all physical phenomena in terms of pure information,[8] and some go so far as to claim that the universe is a gigantic computer. For centuries, scientists and philosophers have conceived of the pinnacle of contemporary technology as a metaphor for nature. The ancient Greeks, inspired by the invention of the ruler, compass, and musical instruments, developed a worldview based on geometry and the music of the spheres. In Renaissance Europe, natural philosophers constructed mechanistic theories inspired by the clock, which represented the finest in contemporary craftsmanship. In the nineteenth century, physicists used the metaphor of the steam engine to explain the universe as a thermodynamic system succumbing to entropy as it winds down to a final heat death. Now the computer is the most sophisticated technology, inspiring neuroscientists to conceive of the brain as a computer, and some physicists use the same metaphor to describe the universe as a whole.

Since computers are constructed piece by piece and the universe evolves through natural processes, perhaps it would be more appropriate to liken the universe to a cosmic brain. Rather than thinking of the universe as matter in motion—as in classical physics—some physicists regard it as information being processed. But whether that information exists in a computer, a brain, or a cosmos, we inevitably come back to the same point: meaningful information exists only relative to the act of informing and a conscious being that is informed. It is not intrinsic to any computational system, so it has no objective existence independent of a conscious subject. Theists may infer from this the existence of a cosmic consciousness, possibly of the kind envisioned by Spinoza and endorsed by Einstein,[9] which pervades all things, is the essence and necessary cause of all things, and comprehends all things.[10] Einstein expressed his own belief in the existence of a "superior mind that reveals itself in the world of experience."[11] But others take a more pluralistic view of the relationship between the mind and nature, an alternative that will be discussed later in this chapter.

Information lies at the core of a participatory universe emerging according to the principle of "it from bit." The observer acquires, records, processes, and replicates information of the semantic kind. For instance, an interaction in quantum mechanics becomes a true measurement only if it *means something* to somebody. This principle in the world of physics is analogous to the hard problem in the world of neuroscience. It is just as difficult to conceive how an abstract notion like meaning or semantic information can emerge from mindless atoms as it is to imagine how consciousness or any other mental phenomenon can emerge from mindless neurons. However challenging it may be to incorporate conscious subjects as active participants in the emergence of the known universe, Wheeler is committed to moving the framework of science onto the foundation of elementary acts of observer participancy.

Such a move requires a profound departure from the metaphysical assumptions of classical physics, in which a property of a system exists prior to and independent of observation, and information is a secondary concept that measures what we learn about properties of the system. In quantum physics, this situation is reversed: the notion of the total information of a system is a primary concept, independent of the experimental procedures chosen by the observer; a property emerges as a secondary concept, namely as a specific representation of the information about the system that results from a measurement. Any observer is able to distinguish a finite number of results at any point in time, and measurements essentially consist of a stream of "yes" or "no" (binary) answers to the questions posed to nature. All concepts of reality based on such measurements consist of mental constructions based on those answers. Those constructs, or theories, are not purely subjective fabrications, nor are they "re-presentations" of an independent, preexisting reality. Rather, they emerge from the interaction between the observer and the observed, and physical objects are identified as having sets of qualities that do not change under various modes of observation or description. These qualities are recognized as invariants, and predictions based on them may be checked by anyone. When those predictions are corroborated by multiple scientists working in multiple laboratories, intersubjective agreement about a scientific model is reached. In accordance with the deeply ingrained human tendency to reify all objects of knowledge—perceptual and conceptual—scientists are naturally prone to attribute a sense of independent reality to these mentally constructed objects.[12]

As Anton Zeilinger comments in this regard, we may be tempted to assume that whenever we ask questions about nature, there is a reality existing independently of what can be said about it. But advances in quantum physics indicate that such a theory is meaningless, and something that is meaningless doesn't even rise to the level of being untrue. Properties we attribute to the objective world can only be based on information we receive from measurements. Apart from such information, nothing can be said about the world of nature that lends itself to either confirmation or repudiation. In other words, natural science is a science of information, not a science of a world that exists prior to and independent of information. So the distinction between information, or knowledge, and reality is meaningless. The collapse of this distinction portends an unprecedented unification of psychology and physics. This may appear difficult, but recall that unification is one of the main themes of the development of modern science, so it is a worthy goal.[13]

## From Quantum Physics to Quantum Cosmology

The emergence of quantum physics from the background theory of classical physics raised fundamental questions about the relationship between scientific theories and the natural world. Einstein succinctly expressed the classical view when he declared, "Physics is an attempt to grasp reality as it is thought independently of its being observed."[14] Over the course of his long-term debate with Einstein, Niels Bohr countered that in quantum physics "any observation necessitates an interference with the course of the phenomena, [and requires] a final renunciation of the classical idea of causality and a radical revision of our attitude towards the problem of physical reality."[15] In quantum physics, unlike classical physics, a subsequent measurement limits the significance of a previous measurement for predicting the future course of phenomena. This implies limits to both the *extent* of the information obtainable by measurements and the *meaning* that can be attributed to it. Bohr concluded therefore that the purpose of devising scientific theories is not to disclose the real essence of phenomena but only to explain relations between the manifold aspects of our experience.[16]

How we interpret the implications of quantum physics with regard to the rest of scientific knowledge depends in large part on our background theory. For most physicists and virtually all biologists and psychologists today, that background theory is classical physics. Most of science consists of knowledge of the past, of real, objective, physical realities, whereas quantum mechanics attends to the future, in which everything is comprised of abstract "waves of probabilities." From the perspective of the background theory of classical physics, every quantum mechanical foreground statement is relative in the sense that it describes possible futures predicted from a particular past-future boundary. Classical descriptions, on the other hand, present the universe as an absolute space-time continuum without distinction between past and future. Consequently, quantum mechanical descriptions are viewed as partial, for they refer only to particular regions of space-time separated from the rest of nature, which is described in classical terms.

The above interpretation of quantum physics in relation to classical physics looks to the past—to the worldview of nineteenth-century science—to interpret the future. Physicists cited in this volume, including John Wheeler and Anton Zeilinger, look to the future by taking quantum physics as their background theory from which to interpret the classical physics of the past. This debate is one of innumerable instances in the history of science of underdetermination, the thesis that empirical evidence alone is never sufficient for proving the truth of a theory. One reason is that foreground theories are always translated and interpreted in terms of background theories, and since there are always multiple background theories to choose from, the truth of a theory is never determined solely on the basis of evidence.[17]

The most striking and mysterious feature of quantum physics is the necessity to draw a line between the observer-participant and the system under investigation. This is the clue to the quantum mechanical construction of everything out of nothing, which is to say, to the manifestation of classical physical realities out of quantum probabilities. Most quantum physicists, following the predominant Copenhagen school, believe that the role of the observer in quantum mechanics is to cause an abrupt "reduction of the wave-packet," or narrowing of the probabilities to a single choice, in which the state of the system appears to jump discontinuously at the instant when it is observed. But others argue that what really happens is that the quantum mechanical description of an event ceases to be meaningful as the observer's point of reference changes from before the event to after it. According to this view, all that is needed to make quantum mechanics work is a point of reference to separate past from future and thereby separate facts from probabilities.[18] A human observer is not necessary, but there must still be some kind of participant who determines the cognitive frame of reference.

A central theme advocated by John Wheeler is that "no elementary phenomenon is a phenomenon until it is a registered (observed) phenomenon."[19] But neither the act of measurement nor the consciousness of that act causes any physical entity to come into or go out of existence. It is also wrong, he maintains, to think of the past as already existing independently of any research protocol, for the past has no existence except as we record it in the present. The answers we get depend on the questions we pose, the experimental designs we create, and the registering devices we choose. In deciding on those, we have undeniable choices regarding what can be said about the past. Likewise, from the perspective of future researchers, the interpretation of quantum events in our present might depend on decisions made in the future concerning measurements performed at some distant space-time location. As Einstein remarked, the very notion of "now" has no objective meaning in the spatially extended world, for "time and space are modes by which we think and not conditions in which we live."[20] This is a central theme in Wheeler's vision of a participatory universe.

One of the most persistent and unresolved problems in the history of theoretical physics is the question of how the quantum and classical worlds are related to each other. The dominant view among physicists today is that quantum systems do not exist in isolation. Quantum wave packets retain their relative phases only until they interact with the environment, which immediately scrambles the phases up. This interaction, called "decoherence," dissolves all the weirdness of quantum phenomena and is crucial in generating a classical world from its quantum components. The theory of decoherence provides the primary justification for confining quantum mechanics to a very small domain of physics. But there is one problem with this explanation: what if there is no environment with which quantum phenomena can interact? The only way that would be possible is if the quantum system under investigation is the whole universe. This possibility inspired Wheeler and Bruce DeWitt to apply quantum mechanics, a theory of the subatomic realm, to cosmology, thereby creating the field of quantum cosmology.

The history of science is characterized by the foreground theories of the present gradually turning into background theories for future scientists. Rather than taking classical physics as the background theory for quantum mechanics, as was done by both Einstein and Bohr, Wheeler presents quantum physics as the background theory for reappraising all of physics, from elementary particle physics to cosmology. One of his points of departure has to do with the so-called "delayed-choice experiment," which he first proposed in 1978. Such an experiment is designed so that the choice of whether one detects a particle or a wave is delayed until *after* the experimental setup has determined that a photon will manifest as one or the other.[21] The experimenter can participate not only in the nature of physical reality that *is* but also in the nature of the physical reality that *was*. Before the experimenter decides how to make the measurement, the light is neither wave nor particle. Only when the decision is made is light realized as having either wave or particle qualities. The remarkable thing about the delayed-choice experiment, which has now been conducted in laboratories, is that the light takes on its qualities at a time *before* the decision is made. Although it appears that an action in the present is influencing an event in the past, this is not so: it is impossible for the experimenter either to send information back in time or to cause a physical effect to occur in the past.

In principle, there is no limit to the internal complexity of a system that can show quantum effects. So Wheeler extended the principle of the delayed-choice experiment to light throughout the cosmos. He envisioned the situation of an experimenter on earth today performing a delayed-choice experiment on photons emitted from a galaxy billions of light-years away, and by the choice of his experimental design, here and now that observer participates in concretizing the physical universe at a time before life on earth existed.

In applying quantum principles to the universe at large, Wheeler is not claiming that the physical universe doesn't exist unless it is observed, only that past states are less than real—that is, they do not possess a full set of physical attributes such as a definite position, motion, and so on. Present observers have a participatory role in determining the actuality of even the remote past.[22] In this vision of quantum cosmology, the universe is a self-excited circuit. In Wheeler's own words: "Beginning with the big bang, the universe expands and cools. After eons of dynamic development it gives rise to observership. Acts of observer-participancy—via the mechanism of the delayed-choice experiment—in turn give tangible 'reality' to the universe not only now but back to the beginning. To speak of the universe as a self-excited circuit is to imply once more a participatory universe."[23]

As we observe and interpret measurements of the cosmic background radiation and other evidence of the origins of the universe, we collectively create the big bang and the evolution of the universe from the cognitive frame of reference of twenty-first-century cosmology. We construct a vision of the universe in which the big bang was followed by the formation of elementary particles, forces, atoms, molecules, and so on as we conceive of them today. Wheeler's notion of a participatory universe has been linked to the anthropic principle, which asserts that the universe is the way it is because we are here.[24] This implies that as long as humans view the universe by way of human concepts, which we impose on our experience, we are always engaging with an anthropocentric universe—we are at the center of the universe we inhabit and explore. This is not to say that the universe, including all other conscious beings, did not exist before the emergence of life as we know it, or that it will vanish when the human species becomes extinct. Rather, only the universe as we conceive of it as existing in the past, present, and future will vanish. More generally, all possible worlds vanish simultaneously with the disappearance of the cognitive frames of reference within which they are apprehended. The worlds experienced by other conscious beings will continue to exist relative to them. In this sense, conscious observers cocreate the worlds in which they dwell.[25]

## The Many Worlds of Ontological Relativity

An essential feature of quantum reality is that it includes the physical world as a whole that does not consist of parts. We can describe this undivided world only if we introduce distinctions in terms of cognitive frames of reference: our world "divides into facts" only because we so divide it.[26] While Wheeler and Zeilinger both acknowledge the indispensable role of the observer in quantum mechanics, they also insist that the conscious observation of a measurement does not in any way influence quantum processes.[27] As Wheeler maintains, the physical system of measurement has an undeniable role in bringing about an event that makes itself known by an irreversible act of "amplification" from the domain of quantum physics to the world we perceive as real.

This assertion returns us to the question: What demarcates the two objective elements of the system being measured and the system of measurement, and the subjective element of the conscious experimenter? Scientific inquiry since the time of Descartes has been almost universally based on the assumption that an absolute distinction can be made between the experimenter and the system under study. But in light of contemporary physics, such an absolute split is no more justifiable than drawing a demarcation between time and space.[28] It is hard to imagine how any explanation of the role of the observer-participant in quantum physics can be complete without including that person's mind, but most physicists today still believe that any references to the mind lie essentially outside physics and can therefore be safely disregarded. While this does allow one to successfully solve all practical problems of quantum mechanics, it brings one no closer to understanding the nature of the observer.

One physicist who challenges this absolute, Cartesian separation of physics from psychology is Michael Mensky, who presents an ingenious theory that explicitly includes the observer's consciousness in the theoretical description of quantum measurement. Mensky's theory is an extension of the so-called "many-worlds interpretation" of quantum mechanics, proposed in 1957 by Hugh Everett in his doctoral thesis, developed under his advisor, John Wheeler.[29] Everett called his hypothesis the "relative state interpretation of quantum mechanics," but ever since Wheeler and DeWitt wrote their papers on this theory,[30] it has been known as the many-worlds interpretation. Everett's paper was long ignored by the scientific community, but over the past twenty years it has attracted growing attention and respect, so that today it is regarded as one of several mainstream interpretations of quantum theory.

According to the most common (Copenhagen) interpretation, the act of quantum measurement entails the "collapse of a wave function," or a "reduction," such that among the range of probabilities, one is selected and all the alternatives vanish. But the laws of quantum mechanical systems do not provide for this transition. The so-called "reduction postulate" was introduced into the quantum theory of measurement to describe what is actually observed in the measurements of quantum systems using classical methods. This description artificially imposes classical concepts upon quantum theory so as to allow scientists to make practical calculations. But it evades the conceptual problems of quantum measurement without really solving them. If we assume that quantum theory is correct and not in need of this intrusion of classical concepts, then this reduction cannot take place at all and should be kept out of the theory.

According to Everett's hypothesis, no such artificially introduced reduction, or selection of a single alternative, occurs. Instead, the act of measurement divides the quantum world into alternative classical worlds, all equally "real." The observer subjectively perceives only one classical world, namely that of everyday experience. But in reality, in all the unseen alternative worlds it is as if replicas of the observer exist, whose experiences provide each of them with a picture of precisely the world they believe they inhabit.

Mensky extends Everett's theory by addressing the question: What happens to the observer's consciousness when such a measurement takes place?[31] Because every observer sees only one measurement result, it seems that a reduction must occur in that person's consciousness, resulting in the choice of one alternative reality out of all possibilities. Mensky overcomes this apparent contradiction in the many-worlds theory by suggesting that "All alternatives are realized, and the observer's consciousness splits between all the alternatives. At the same time, the individual consciousness of the observer subjectively perceives what is going on in such a way as if there exists only one alternative, the one she exists in. In other words, *the consciousness as a whole splits between the alternatives but the individual consciousness subjectively chooses (selects) one alternative.*"[32] This implies a kind of symmetry between what happens to the experimenter and what happens to the system under study. Consciousness does not mechanically cause the wave function to collapse or influence physical particles. Rather, the observer's brain and the observed system are synchronously entangled. "Entanglement" is a term used in quantum physics to describe the nonlocal connectedness of two or more phenomena, which exist as one system even if they are separated by a very large distance.

With classical physics as the background theory, the quantum world is regarded as a set of mathematical abstractions in the form of probabilities with which to predict classical reality, which is therefore viewed as objective. But in Mensky's hypothesis the relationship between the quantum and the classical worlds is reversed: quantum reality, consisting of parallel worlds, all equally real, is considered to be objective because it does not depend on the consciousness of the individual observer. Each classical world is just one more "classical projection" of the quantum world that comes into existence only when the observer's consciousness selects one of the parallel worlds. This classical world is an illusion, for it exists only relative to the mind of the observer. Quantum physics is thus fully established as the background theory for the foreground theory of classical physics.

Mensky's theory also extends earlier interpretations of the anthropic principle. Unlike inanimate matter, every conscious living being perceives the quantum world, with its characteristic nonlocality, relative to its own cognitive frame of reference. Each of these individual classical projections is "locally predictable," and in each one, a conscious being realizes a world of lived experience. And each such classical world exists only relative to such a being or community of beings.

It is important to note that in any one of Everett's worlds, because of the internal principles of quantum mechanical evolution, all valid observers within the same cognitive frame of reference see the same thing, so their observations are consistent. The selection of reality by conscious observers is made in relation to those aspects of reality they consider most vital. So it is possible that by choosing a classical world according to our interests, we may affect the probability of which alternative we observe. Mensky explains: "If, for instance, a close relative dies in one of these realities and remains alive in another, the conscious subject is highly motivated to select the latter alternative. If he believes in this case that he is able to affect the selection, it is not inconceivable that he will actually increase the probability to some extent that he will witness precisely the latter alternative."[33]

This hypothesis raises the possibility that individuals may alter the course of events by their choices, aspirations, faith, and prayers. This may help to explain the mysterious "placebo effect" familiar from medical studies, and it may even introduce the possibility of "miracles" into contemporary physics. But even in the case of alleged miracles brought about, for example, by the power of prayer, Mensky cautions that individuals could never be certain that they were the ones who affected the course of events. Even if their beliefs, thoughts, or prayers repeatedly corresponded to what happened, skeptics could always counter that events could have taken this course in a "natural way," in accordance with the mindless laws of classical physics. Even if such a skeptic were to personally witness a "miracle," there would still be room for doubt, but the likelihood of such an observation would be decreased by the fact that skeptics prefer to live in a world where such paranormal events are impossible. They will remain in their cognitive frame of reference in those of Everett's worlds where ordinary physical laws rule the objective universe without interference from subjective influences—or at least, so they may believe. But those who believe in this hypothesis have ample opportunities for tapping into multiple worlds where such "probabilistic wonders" do occur.

Mensky's extension of Everett's theory has two possible implications: this theory cannot be incorporated into the realm of science, or scientific methodologies have to be radically broadened to include observations and experiments that accept the mind as a cocreator of the natural world. Such new methodologies would have to include "first-person science," involving experiments in individual consciousness as the instrument of theory verification, and a consideration of the possible effects of the investigator's background theory and preferences on the results. Even if this theory is accepted as a working hypothesis, the objectivity of scientific results is guaranteed within the cognitive framework of the conventional scientific community, which internally shares its discoveries without seeking to alter them by selecting from "nonscientific" alternative realities. So conventional science is secure and can continue to evolve as if the mind had no active role in nature.

I shall conclude this chapter in Mensky's own words:

Work in the context of Everett's concept calls for the extension of methodology and in some sense leads out of the province of physics and even natural sciences in general. This, of course, should cause anxiety and raise debate. However, the problem seems to be extremely important, which justifies even speculative steps. In the event of success in the solution of conceptual problems of quantum mechanics, this science, as well as physics entirely, is certain to rise to a qualitatively new level of the understanding of nature. If we are guided by what has already been done in the framework of Everett's concept, physics would be expected to form a fruitful symbiosis with psychology, as well as with other means of cognition of spiritual human life.[34]

# 8

# EXPERIMENTS IN
# QUANTUM CONSCIOUSNESS

## Parallels in Quantum Physics

In the autumn of 1997, a small group of distinguished physicists, astronomers, and philosophers gathered for five days at the Dalai Lama's home in the foothills of the Himalayas in northern India to discuss the interface of quantum physics, cosmology, and Buddhism.[1] Anton Zeilinger, one of the world's foremost experts in the experimental foundations of quantum mechanics, was a prominent participant. He is best known for his groundbreaking experiments at the University of Innsbruck that demonstrate quantum teleportation, or the transmission of an exact replica of an arbitrary quantum state to a distant location.[2] During this meeting, Zeilinger explained to the Dalai Lama the wave-particle duality for single photons, the concept of objective randomness in quantum mechanics, and the profound mystery of nonlocality. To illustrate some of the strangeness of quantum physics, he even brought with him a miniaturized system of quantum measurement.

Over the course of his dialogues with the Dalai Lama, Zeilinger became intrigued by the fact that Buddhist philosophers and contemplatives, without knowing anything about modern physics, had concluded that no phenomenon has inherent, objective existence, independent of the means by which it is apprehended. The Dalai Lama was equally fascinated that quantum physicists, without knowing anything about Buddhist philosophy or meditation, could have come to a similar conclusion. This lively encounter led to the Dalai Lama accepting an invitation to visit Zeilinger's laboratory at the University of Innsbruck the following summer, where he was shown various experimental procedures in quantum physics in greater detail. The discussions of modern physics, philosophy, and Buddhism that ensued during this visit were deeply stimulating and provocative for everyone involved.

In all Zeilinger's comments during these meetings, as in his experimental research generally, he tried to stay as close as possible to the phenomena of quantum measurements, using a minimum number of presuppositions. In adopting this stance, he follows in the footsteps of Isaac Newton, who declared in his *Mathematical Principles of Natural Philosophy*, "I feign no hypotheses." Newton made this comment specifically with regard to gravity, contenting himself with setting forth its mathematical principles, not speculating on its actual nature. Zeilinger, like Newton, avoids "hypotheses" in the sense of propositions that are only assumed without being deduced from experimental evidence. Following his remarkable success in demonstrating quantum teleportation, he was appointed professor of physics at the University of Vienna, where he holds the position previously occupied by Erwin Schrödinger.[3]

For all the ingenuity of recent experimental research in the foundations of quantum mechanics, the role of the observer remains as mysterious as ever, and the demarcation between the system of measurement—which is widely viewed as a classical system—and the quantum system being measured is simply accepted as a given. The role of the subject remains outside contemporary physics, as it was fifty years ago, when Schrödinger wrote: "Without being aware of it, we exclude the Subject of Cognizance from the domain of nature that we endeavour to understand. We step with our own person back into the part of an onlooker who does not belong to the world, which by this very process becomes an objective world."[4]

Given the radically different methods of inquiry used in physics and Buddhism, it is remarkable that they both present visions of the natural world in which all phenomena manifest as quanta, or discrete units. In physics an atom, having a diameter of approximately $10^{-8}$ cm, is regarded as the smallest particle of matter that cannot be taken apart by chemical means. On a more fundamental level, all configurations of mass-energy consist of quanta, which in the case of light are called photons. Mass is a property of a physical object that quantifies the amount of matter and energy to which it is equivalent. Just as mass can be completely transformed into energy, energy can be totally transformed into matter. Energy is always a property of something else, and it has no existence apart from matter. They differ only in terms of the units in which they are measured.

In Buddhism also, an atom (*paramaṇu*) is the smallest physically indivisible unit of matter, and traditional sources claim that it has a diameter of $10^{-9} - 10^{-10}$ cm.[5] Buddhist physics describes the world as consisting of five elements, discussed in chapter 5—solidity (earth), fluidity (water), heat (fire), motility (air), and space—and there is a definite sequence in their emergence. According to Buddhist cosmogony, our universe evolves and devolves in an oscillating cycle, compatible with the scientific idea of multiple big bangs. In the evolution of the universe, motility, or kinetic energy, emerges from space; heat, or thermal energy, emerges from motility; fluidity emerges from heat; and solids emerge from fluids. Thus, each of the latter elements is an emergent property of the former, and all are derivative of space. In the eventual destruction of the cosmos, all the derivative elements dissolve back into space. This implies that all forms of matter are thoroughly interchangeable with kinetic energy, and all forms of matter and energy are transferable into space.[6] One fundamental difference between scientific and Buddhist views of the universe is that science traditionally seeks to describe the physical world as it exists independent of any observer, whereas Buddhism is concerned only with the world of experience (*loka*), which is inseparable from conscious subjects. The importance of this distinction cannot be overemphasized.

Early in the twentieth century, when quantum theory was first formulated, physicists speculated about the possibility of a quantum of time, which they called a chronon. This is the shortest duration of a single identifiable change, and it is *very* roughly calculated to be $10^{-24} – 10^{36}$ seconds.[7] Some physicists believe that the continuum of space-time is likewise made up of discrete units, and in order to account for its continuous symmetry, these units should be understood in terms of quantum theory, not classical physics. But unlike in mass-energy, there is no symmetry between time and space. According to the mathematical theory of Einstein's relativity, space and time transform to a limited degree. Inside a black hole, space is completely transmuted into time, which is to say it enters into time but doesn't actually become time. Likewise, time enters into space, but it doesn't become space.

All schools of Buddhism also regard time as consisting of discrete units, though they disagree on the exact duration of a Buddhist chronon (*kṣaṇa*), also identified as the smallest unit of time in which change can occur. Some treatises declare this to be on the order of $10^{-3}$ seconds, and they state that the briefest mental phenomena undergo changes 16 to 17 times as fast as the shortest changes in the physical world. But other sources say that billions of mind-moments arise and pass within the duration of a flash of lightning. They add that time does not exist by its own inherent nature, but only as conceptual imputation based on other phenomena.[8] So the very notion of time is meaningful only in relation to a specific cognitive frame of reference; it has no objective existence independent of any observer-participant. The Buddha himself would not specify the precise duration of the shortest unit of time, saying that no one is capable of understanding it.[9] All Buddhist schools agree, however, that the smallest units of matter and time are not detectable with ordinary perception, only with the heightened awareness of contemplative perception (*yoga-pratyakṣa*).[10]

Scientific theories of the elementary units of information, or bits, were discussed in the previous chapter. Buddhism also regards information as being comprised of units of syllables (*akṣara*), which are the elementary building blocks of words. As we shall see later in this and the next chapter, words and concepts play a fundamental role in the formation of the known universe in Buddhism, as they do in Wheeler's concept of the participatory universe as a self-excited circuit.

## Exploring the Quantum World of Experience

The above Buddhist accounts of the quantization of matter, energy, space, time, and consciousness all raise the question: How did Buddhists come up with these theories? Are they simply metaphysical speculations or precise reports of contemplative experience? In most Buddhist writings, no clear demarcation is drawn among experiential reports, theoretical inferences, and philosophical speculation, so each has to be evaluated on a case-by-case basis. These theories are to be treated like any other scientific ideas, by testing them experientially, rather than judging them in terms of prior assumptions. But to put these Buddhist hypotheses to the test of experience requires years of professional training, as would testing any other sophisticated scientific theory.

Insofar as any of the above assertions is based on immediate experience, such contemplative insight (*vipaśyanā*) has as its minimum prerequisite the accomplishment of a highly refined degree of focused attention known as meditative quiescence (*śamatha*).[11] Quiescence is to contemplative discoveries what the telescope is to astronomical discoveries, and any meditator who has not yet achieved it is technically regarded as a novice.[12] The previously described practice of settling the mind in its natural state culminates in quiescence, initially gaining access to the form realm by way of the substrate consciousness. Once one has achieved this exceptional level of attentional balance, one should be able to effortlessly remain there, with the physical senses totally withdrawn, for at least four hours, with unwavering mindfulness and an extraordinary degree of vividness.

In addition to remarkable state effects, quiescence also yields a number of trait effects that carry over into daily life while not formally engaged in meditation.[13] Generally speaking, mental imbalances such as craving and anger arise only infrequently, and when they do occur they are feeble and of brief duration. We experience an unprecedented degree of mental and physical suppleness and fitness, and the qualities of attentional stability and vividness persist even in sleep. Our attention remains highly focused throughout all activities, and there is such high resolution and vividness in our awareness that we feel as if we could count the individual atoms of the material objects in our environment. Moreover, visual experience of all kinds of physical phenomena may be radically transformed, so that we perceive them simply as matrices of luminous forms rather than as concrete objects.[14]

The earliest records of the Buddha's teachings emphasize the importance of quiescence, which is technically identified as access to the first meditative stabilization (*dhyāna*) in the form realm.[15] The indispensability of quiescence for the cultivation of contemplative insight is reiterated in the later development of Mahāyāna and Vajrayāna Buddhism in India and Tibet.[16] In a number of his discourses, the Buddha, like Newton and Zeilinger, encouraged his students to attend closely to phenomena without superimposing their preconceptions onto the immediacy of experience.[17] But this does not imply the mere cultivation of "bare attention" in the sense of mindfulness completely devoid of a theoretical framework, ethical evaluation, or purposive direction. Nowhere in any of the hundreds of recorded discourses of the Buddha does he ever equate mindfulness with bare attention. Authentic mindfulness (*samyak-smṛti*) is practiced in conjunction with rigorous analysis (*samyak-saṃkalpa*) and theorizing (*samyag-dṛṣṭi*). The notion of bare attention is closer to the more primitive mental factor of mental engagement (*manas-kāra*), which has the function of directing the mind to an object or selecting features of the object for close attention. The Buddhist concept of mindfulness, in contrast, has a strong ethical component, and its primary characteristics are unwavering, penetrating, discerning attentiveness to the object under examination.[18]

The most fundamental system of all Buddhist contemplative insight practice is known as the Four Applications of Mindfulness, in which one meticulously scrutinizes the body; psychophysiological feelings of pleasure, pain, and indifference; mental states and processes; and phenomena at large.[19] Such inquiry entails a unified pursuit of genuine happiness, understanding, and virtue, and the Buddha declared it to be a "direct path" to freedom from suffering and its inner causes. The Buddha instructed that one should examine one's own body internally, consider others' bodies externally, and contemplate one's own and others' bodies both internally and externally. Continuing to scrutinize the body, one examines the factors giving rise to its origination, its dissolution, and both its origination and its dissolution. Finally, one simply sustains mindfulness of the sheer presence of the body to the extent necessary for apprehending its nature.[20] These same methods are then applied to the other three subjects of inquiry.

With attention already refined by means of the achievement of meditative quiescence, one inspects all the above classes of phenomena to determine whether they are of a stable, enduring nature or are momentary and impermanent; by nature satisfying or unsatisfying; and by nature "I" or lacking such personal identity. One Buddhist hypothesis is that all phenomena that arise in dependence upon causes and conditions are of a transient nature, and they naturally disintegrate from moment to moment even without the influence of external factors. As long as one's mind is subject to such afflictive tendencies as delusion, craving, and hostility, it is said that no objects or subjective mental states will prove ultimately satisfying. Finally, all phenomena, including one's own body and mind, are declared to be devoid of an enduring, unitary, independent self. A general challenge of this entire discipline of inquiry is to distinguish between the phenomena that are actually perceived and the conceptual superimpositions we project on them.[21]

In short, as the Buddha counseled, one seeks to come to the realization that "In the seen there is only the seen; in the heard, there is only the heard; in the sensed, there is only the sensed; in the cognized, there is only the cognized."[22] By perceiving the immediate contents of experience, stripped as much as possible of unwarranted assumptions, one realizes the sheer absence of a separate, autonomous self, either internal or external.[23] More generally, one realizes that *things* have no independent existence, in the outer world, the inner world, or anywhere in between. Just as the label for an external entity such as a chariot is applied to a configuration of parts, so is the name of a person imputed upon a body and mind. But prior to and independent of these conceptual projections, these entities do not exist in and of themselves. The Buddha declared, "These are merely names, expressions, turns of speech, designations in common use in the world. Of these he who has won truth makes use indeed, but is not led astray by them."[24]

## Exploring the Illusory Worlds of Dreaming and Waking

The practice of the Four Applications of Mindfulness yields insight into the illusory, dreamlike nature of reality during the waking state. We may also continue this mindful examination of phenomena during the dream state so as to more deeply understand the similarities and differences between waking and dreaming realities. This requires learning to recognize that we are dreaming while we are dreaming, for which it is necessary to sustain a strong resolution throughout the day to retain mindful awareness in the dream state. Then we must have dreams and achieve sufficient clarity in them so that we can describe them in detail to others. For this, dream journaling can be very helpful. It may be necessary to withdraw temporarily into solitude, which enhances the clarity of dreams, and then to focus on the dreams that occur in the early morning, when they are especially clear.[25]

There are various catalysts that may enable one to become "lucid" during the dream state, that is, to recognize it for what it is. Some people recognize that they are dreaming due to the power of their resolve during the daytime. Others are brought to a state of lucidity by recognizing an anomaly in a dream, something that couldn't possibly happen during the waking state. And yet others fall asleep without losing mindfulness, and as soon as a dream appears they recognize it as such. It is even possible to retain discerning mindfulness during dreamless sleep, enabling one to explore the nature of the substrate consciousness. If we have already achieved meditative quiescence, we will find it quite easy to become lucid in our dreams and to sustain such mindful awareness, and can then explore the general and specific characteristics of dream phenomena.[26] In this way, we may continue in the practice of contemplative insight during all states of consciousness, while awake, while dreaming, and in dreamless sleep.

The traditional Tibetan Buddhist practice of dream yoga involves four trainings: learning to maintain lucidity during dreams, regulating the qualities and quantities of dream phenomena, overcoming fear by recognizing the illusory nature of dreams, and meditating on the actual nature of dream phenomena.[27] Both the modern discipline of lucid dreaming and the ancient tradition of dream yoga offer a wide range of practical techniques for recognizing the dream state and sustaining that awareness with stability and vividness. The prior accomplishment of meditative quiescence is good preparation for the first phase of this contemplative training. In the second phase, we begin to explore the nature of dream reality by deliberately altering the contents of dreams. For example, we may voluntarily generate a specific dreamscape or series of events or transform the contents of a dream that has already arisen. We may transform our own appearance, as well as anything else, and reduce or increase the quantity of anything in the dream. In these ways we discover whether there is anything in the dream, "subjectively" or "objectively," that is impervious to our will and the powers of our imagination. In other words, we train in selecting the reality of our choice from a myriad of possible worlds, and this is done by the power of conceptual designation and imagination.

The above practices of dream yoga are bound to catalyze occasional terrifying and traumatic dreams, such as drowning in water or being burned by fire. In the third phase of this training, whenever anything of that sort happens, we recognize the dream as such and ask ourselves, "How can dream water or dream fire possibly harm dreamed me?" With the clear recognition that everything in a dream consists of mere appearances, no more substantial than a rainbow, we can jump into the water or fire. Our own form in the dream is an illusion, as is the water or fire; even if we perish in the dream, that too is an illusion. With that clear recognition, we cannot be harmed, so there is no need to attempt to escape harm by transforming or terminating the dream.

The fourth practice of dream yoga entails thoroughly fathoming the illusory nature of everything in the dream, fearful or otherwise. Now we realize that nothing in the dream exists by its own inherent nature, independently of the perceptual and conceptual framework in which it is apprehended. All objective and subjective dream phenomena are "empty," yet in dependence upon prior causes and contributing conditions, they are produced and in turn give rise to and influence other phenomena in the dream and in waking reality.

In addition to the above four practices, we may allow a dreamscape to dissolve back into the empty space of the mind. This can easily be done simply by being still in the dream. Ceasing to interact with dream phenomena and suspending all conceptual designations upon dream events allows them to quickly disappear. All that remains is a vacuous space of awareness, devoid of thoughts and images. This is the substrate, out of which all dream phenomena appear and into which they dissolve. The substrate is most clearly apprehended in the state of quiescence, but we can also experience it in lucid, dreamless sleep. We may then examine the way dreams emerge from this mental vacuum, how they consist of nothing more than configurations of mental space, and how they eventually dissolve back into that space.

When we wake up, dream appearances vanish into the substrate and waking appearances emerge from that same substrate. The primary difference between them is that dream phenomena are not directly dependent upon physical influences from the surrounding environment, whereas waking phenomena are. Nevertheless, even the physical phenomena we experience during the waking state do not exist independently of our cognitive frame of reference. When we fall asleep, perceptual and conceptual appearances dissolve back into the substrate, from which dreams eventually emerge once again. So the daytime practice of dream yoga has two central themes: recognizing how all phenomena in the waking state are momentary and exist only relative to perceptual and conceptual frameworks, and recognizing that even though all phenomena are empty of their own inherent nature, they still appear and causally interact with each other in lawlike ways.[28] Physical phenomena, for instance, are relative and illusory, yet they still arise and interact according to the laws of physics. And even though there is no inherently existing self, or ego, people still engage in actions and experience the results of their behavior in lawlike ways. In science, the laws of nature are not simply a matter of personal preference, and in Buddhism, the laws of karma are invariant across multiple cognitive frames of reference.

## Exploring the World of Quantum Relativity

Buddhism is one of many contemplative traditions that emphasize the participatory nature of reality and the illusory nature of appearances. Erwin Schrödinger, Anton Zeilinger's predecessor at the University of Vienna, was a student of Sanskrit and Indian philosophy, and he may have been intrigued by the fact that a common Sanskrit term referring to the world of appearances is *māyā*, literally meaning "illusion." Interestingly, this word stems from the Sanskrit verbal root *māya*, which has the dual meaning of "to measure" and "to create illusions." Like quantum physicists, Indian contemplatives connected the act of measurement to the manifestation of illusory appearances.

As Buddhism made its way to Tibet in the eighth century, its central themes were incorporated into the language of this Himalayan culture. One commonly used Tibetan word for "world" (*srid pa*) has the connotation of "possibility" and "the process of becoming"; another closely related word (*snang srid*) refers to "all phenomena that can possibly appear." Returning to the English language, the word "meditation" comes from the Indo-European verbal base "med-" (also related to the Greek term "metron"), which means "measure" or "consider." The recurrent theme here is that the act of meditation, or measurement, divides up the seamless fabric of reality, giving form to manifold worlds of illusory, dreamlike appearances.

According to Buddhist cosmogony, the world of conditioned existence (*bhava*) consists of multiple implicate and explicate orders. Within the context of these "worlds of becoming," the formless realm consists of four dimensions of existence, the subtlest of which are made up of quanta of consciousness, time, and energy (*prāṇa*). From those implicate dimensions emerges the fourth, formless dimension, which also includes space, and from that dimension emerges the form realm discussed earlier. Finally, the explicate order of the physical world—consisting of quanta of consciousness, time, energy, space, and matter—emerges from the form realm. During the course of cosmic evolution, each of these dimensions of reality emerges from its respective implicate order, and according to the Middle Way (Madhyamaka) view of Mahāyāna Buddhism, all of them exist only in relation to their respective cognitive frames of reference.

To put this theory to the test of experience, we may begin by investigating the nature of the mind.[29] First of all, to examine the nature of consciousness itself, we let our awareness come to rest in the space in front of us, and without focusing on anything, simply maintain unwavering mindfulness of being conscious. When our mind stabilizes in this practice, we examine the nature of that which has become stable. We ask yourselves, "Is there something there that has become still?" Can we find anything that corresponds to the word "mind"? Is the mind nothing or is it something? When we observe our mind, is there one entity that is being observed and another that is doing the observing? To the best of our ability, we invert our awareness upon the observer. As we consider, "What is the nature of the mind?" we direct our attention to that which is posing the question.

If the mind does exist, it must have its own salient characteristics. We examine whether it has any shape or form and whether it is located anywhere in space or has spatial dimensions. If we look for the mind and fail to identify it, this doesn't necessarily imply that it is nonexistent. If we come to that conclusion, we examine that which concludes that the mind doesn't exist. When discursive thoughts, mental images, emotions, desires, and other mental phenomena arise, we investigate how they come into being and where they come from. We examine where, if anywhere, they are located, and when they vanish, inspect precisely how they dissolve and what they disappear into. A major theme of this contemplative insight practice is to examine whether all these mental phenomena are the same or different and to seek out the nature of that which is observing them. A central insight to be gleaned is that all mental phenomena, as well as the observer of those phenomena, are empty of inherent nature. That is, they do not exist independently of any cognitive frame of reference. Not even consciousness is an absolute, for within the world of conditioned existence, moments of subjective consciousness always arise in relation to objects of awareness.

Closely related to this ontological probe into the nature of the mind is the investigation of the self as an agent.[30] In this practice we first examine how we actually experience ourselves as individuals, our sense of personal identity. We see if we regard ourselves as existing by our own inherent nature, independent of any conceptual designation. If we do, we identify the object we grasp onto as being ourself and sustain it, as if we were suspending a particle within a force field. Then we subject that reified self to ontological analysis. We examine precisely how it arises and passes, and while it is present, investigate whether it is identical to or distinct from our body and mind. We may discover that it is neither the same as nor different from our body and mind. In this way we come to realize that such a reified self has no existence whatsoever. This is not a matter simply of failing to find it, but rather of discovering that there is no such inherently existent self at all. This doesn't imply that one doesn't exist as an observer or as an agent, but that we, like our body and mind, arise from moment to moment as a sequence of dependently related events, all emerging in relation to a cognitive frame of reference. So one does exist as much as anything else, and one does engage in actions for which one will experience consequences.

The above methods of contemplative inquiry suggest that no subjective phenomenon has its own inherent or absolute existence. According to the Middle Way, the same holds true for all objective phenomena. Within the context of modern physics, something is said to be objective if, from one observer's description of an entity, the descriptions of all other observers can be computed. In other words, when everyone can translate their descriptions into each other's language, we say they are describing an objective entity. But this means that objectivity boils down to invariants within intersubjective experience.[31] Particularly in quantum physics, when one seeks out the nature of a physical entity as it exists independently of any system of measurement, one does not merely fail to find it. Rather, one discovers that such an independent entity doesn't exist.

Likewise, in the Middle Way analysis of all kinds of physical phenomena, by examining the manner of their origination and by analyzing the relationship between such phenomena and their component parts and attributes, one finds that none of them exists by its own inherent nature.[32] Configurations of mass and energy are not identical to their parts or characteristics, yet they do not exist independently of them. In the dialogues between Anton Zeilinger and the Dalai Lama, both were struck by this extraordinary convergence between quantum physics and Buddhism.[33] As Piet Hut, another physicist at the 1997 meeting, commented, this could be a mere coincidence, but only if the physical world and the mental world are absolutely different without any possibility of transformation. If the themes of relativity and transformation are connected, then their convergence is not accidental. This could imply, he continued, that we are moving from a science of objectivity to a science of intersubjectivity, in which the next relativity theory will include a relativity between the object and the subject, between the physical and the mental.[34]

According to modern physics, the notion of time no longer has any objective meaning. For example, in quantum mechanics two photons may be registered at the same time relative to one location, but somebody else moving relative to that location would detect one photon measurement as occurring before the other. So in that second inertial frame of reference, the first photon would be the cause of the second. Another person going the opposite direction would see the reverse cause and effect. So no event exists absolutely simultaneously with, prior to, or after another.[35] While physics presents time as existing only relative to an inertial frame of reference, Middle Way Buddhism proposes that the passage of time can be measured only relative to a cognitive frame of reference.[36] For example, from the perspective of a mind immersed in a form or formless realm, time passes far more slowly than it does for someone whose mind is engaged with the physical world as we know it. And no unit of time exists independently of its component parts, its beginning and end. The very quantization of time itself is meaningful only relative to its conceptual designation.

In physics, references to time are simply a way to say when something happens, and references to space are a way to say where something is. All references to time and space are meaningful only in relation to something else; in themselves they have no meaning. Points of space are no longer regarded as having any objective existence, and in physics at large, the number of absolutes is steadily decreasing while the range of relative entities is getting larger and larger. One absolute that has survived is Einstein's notion of space-time, to which he attributed independent existence. According to David Finkelstein, another physicist at the above meeting with the Dalai Lama, even without material bodies you can have gravitational waves, which are waves of curvature of space-time, propagating through space-time. Nevertheless, he argued that it is very likely that the idea of an absolute space-time will go the way of many other absolutes.[37] That would certainly accord with the Middle Way analysis of space, which denies its inherent nature as well as that of any possible union of space and time.[38]

The notion of causality and the Buddhist view of dependent origination are both deeply related to time. We have already seen that according to Einstein's theory of relativity, the temporal sequences of causal interactions are always relative to an inertial frame, which refutes any notion of absolute causality.

Zeilinger commented that in quantum physics, the entanglement between two particles is broken when one of them interacts with an outside system such as a detector. In other words, the act of measurement causes the entanglement to cease so that it is no longer present for any future observations. But from another perspective, the detector itself can be viewed as a quantum system, so what takes place is not disentanglement but an increasing entanglement of one quantum system with a much more complex quantum system. According to this view, the whole world becomes entangled.[39] We have already encountered this principle in quantum cosmology, in which no boundary is established between a quantum system and its environment or system of measurement. Zeilinger added that according to the Copenhagen interpretation of quantum mechanics, observation breaks the entanglement, but there is no explanation of what constitutes an observation.

According to the Middle Way view, causal interactions may be sequential, but simultaneous causation may also occur.[40] For example, there can be no actor without an act, so there is a causal interdependence between the two that is simultaneous and mutual. Mental and physical phenomena arise in dependence upon causes and conditions, the component parts and attributes of each entity, and the conceptual designation of those events. But independently of the subjective act of conceptual designation, no objective phenomenon or interaction among phenomena of any kind exists.

The intersubjective nature of the natural world does not imply solipsism in either physics or Buddhism. Laws regulating the interactions among physical phenomena, among mental phenomena, and between physical and mental phenomena can be discovered that are invariant across multiple cognitive frames of reference. But as David Finkelstein notes, such laws can take on the role of idols in the sense of the term discussed earlier in chapter 1. That is, in classical physics a law is expressed as an equation that is completely known and influences what happens in the world. There is no counterinfluence, so in the relation between natural laws and nature, the laws are idols. Throughout the history of physics, the presence of any kind of idol has always been a sign of a degenerate theory, concealing a deeper, underlying symmetry. So, as John Wheeler has proposed, it is possible that laws evolve and are subject to change by what happens in the world. Finkelstein concludes, "The laws of nature affect matter; therefore, matter must affect the laws of nature."[41] Buddhism also maintains that the intersubjective regularities between actions and their long-term consequences, or the laws of karma, shift as one evolves along the path to enlightenment.[42]

In short, contemporary physics presents space, time, mass, and energy as being comprised of discrete units, and for the most part, sees all such natural phenomena as relative to something outside themselves. The Middle Way view, when couched within the larger framework of Vajrayāna Buddhism, similarly regards all subjective and objective phenomena, including consciousness, space, time, energy, and matter, as being made up of discrete units, none of which bears absolute existence. This is the Buddhist general theory of ontological relativity.

## The Union of Quiescence and Insight

As a result of their penetrating investigations of the natural world, quantum physicists and Buddhist contemplatives have both come to challenge the objective, independent nature of the subjects of their inquiry. But the methods by which they have drawn their conclusions could hardly be more different. Physicists rely on third-person observations and experiments, from which the resulting empirical evidence is quantitatively analyzed. Buddhists rely on first-person observations and experiments on mental and physical phenomena in conjunction with qualitative analysis. In Buddhism, preparation for the cultivation of contemplative insight strongly emphasizes the development of ethical discipline and mental balance, especially by developing meditative quiescence.[43] In science, ethics plays a relatively minor role, and there are no practices for the cultivation of highly focused, stable attention or other aspects of mental balance.

As a result of their insights, quantum physicists are inspired to challenge all their assumptions regarding the nature of reality, and this in turn motivates some of them to probe more deeply into what underlies the strangeness of quantum phenomena.[44] Earlier in this volume I proposed three criteria for evaluating theories and methods of human inquiry: in terms of the extent to which they enhance genuine happiness, understanding, and virtue. According to these humanistic criteria, the theories and methods of quantum physics score highly in terms of extending human understanding of nature. They have also significantly contributed to advances in technology, including medical technology, which support our physical well-being. But they have done little if anything to enhance mental well-being or the cultivation of human virtues. Buddhist insight into the emptiness of inherent existence of all phenomena has a profound effect on the mental well-being, worldviews, and virtues of practitioners, and these people exert a powerful influence on their society. So the theories and methods of contemplative inquiry score highly in terms of their contributions to genuine happiness, understanding, and virtue. But they have contributed little to the natural sciences or technology.

Beginning with contemplative insight into the nature of the mind, one discovers that all phenomena arise only in relation to the mind that apprehends them, yet that mind is empty of its own independent, inherent identity.[45] From moment to moment, mental phenomena emerge and vanish only in relation to a cognitive framework. They have no absolute, subjective origination, presence, or dissolution, and neither do all their objective appearances. Fully integrating these insights into everyday life helps to shift priorities from the pursuit of mundane, hedonic pleasures to the pursuit of genuine happiness and freedom from all mental afflictions. Craving, hostility, and delusion all radically subside, and as a result of insight into the profound interdependence of all beings, love and compassion spontaneously arise as never before.

The above benefits of contemplative insight into the nature of emptiness and dependent origination arise most fully from the unification of quiescence and insight. This contemplative inquiry begins with meticulous observation of the phenomenon in question, followed by conceptual analysis of its mode of existence. At first, the resulting insights are purely conceptual, much like those in quantum physics. Indirect, conceptual understanding dispels our previous conceptual errors. Then we apply the stability and vividness of meditative quiescence to the conceptual insights, saturating our mind with these newfound discoveries. As a result of such sustained, experiential familiarization, the veils of conceptuality gradually lift, giving way to direct, nonconceptual realization of the empty nature of phenomena. In this way, conceptual insight counteracts conceptual error, until eventually concepts of both kinds fade away. In one Mahāyāna discourse, the Buddha explained this process as being like two branches igniting from the friction of rubbing together, resulting in the eventual incineration of both.[46]

Due to the unification of quiescence and insight into the nature of emptiness, Buddhists claim that people may develop a wide range of paranormal abilities. The Dalai Lama comments that such contemplatives can experience eons shrunk into a single instant of time and stretch a single instant of time into an eon. That which is experienced as an eon from the cognitive framework of the contemplative would appear as a single instant to a noncontemplative. Although this expansion and contraction of time is unique to the subjective experience of the contemplative,[47] not all paranormal abilities resulting from the union of quiescence and insight are confined to subjective experience. According to many Buddhist accounts, one may mentally alter physical reality in a myriad of ways and gain various kinds of extrasensory perception.[48]

From the perspective of modern science, all such claims must be viewed as uncorroborated hypotheses. But they have not been put to the test of experience because advanced contemplative training has never been part of scientific education or research. Until such practices are incorporated into scientific inquiry, we can at least try to make sense of these claims in light of contemporary physics. For they are utterly implausible in terms of the assumptions of classical physics on which neurobiology and psychology are currently based.

Let us begin with the analogy of lucid dreaming and the practices of dream yoga mentioned above. When we first become lucid in a dream, dispelling the implicit error that we are experiencing waking reality, a sense of euphoria commonly arises. This is not evoked by any particular object or event in the dream, but rather by insight into the nature of our current experience. A much deeper sense of bliss and mental pliancy is said to result from direct insight into emptiness during the waking state, far surpassing even the bliss of achieving meditative quiescence alone.[49] The euphoria of a lucid dream results from insight into the nature of the dream; the joy of quiescence stems from temporary freedom from the attentional imbalances of laxity and excitation; and the bliss of contemplative insight is a result of overcoming the deep-rooted delusion of reifying phenomena of all kinds.

In a lucid dream, we can perform feats that appear miraculous to others in the dream. Simply by imagining that water can support our weight, we can walk on it, just as we walk through walls and fly through the air. By the power of conceptual designation based on the recognition that everything we are experiencing exists only relative to our mind, we can transform our own appearance at will and instantly shift from one dream environment to another. What we are altering is the appearances that emerge from our own substrate, and these influence everyone in our dream, for they also arise from the substrate. Although these transformations may be confirmed by "third persons" within the dream, they are all confined to the individual consciousness of the dreamer. They have no impact on the physical environment of the person who is asleep in bed. In other words, the dreamer lying in bed and the persona appearing in that individual's dream are not the same, for the latter emerges from the substrate of the former. And when the lucid dreamer dissolves an entire dreamscape back into the substrate, all its contents disappear only relative to that dreamer's cognitive frame of reference. Moreover, the lucid realization of the empty nature of dream events does not imply any realization of the empty nature of events in the waking state, so it doesn't have a comparable impact on one's life as a whole.[50]

Likewise, resting in a state of consciousness in which the tendency toward reification is temporarily suspended does not result in any of the benefits of the union of quiescence and insight into the ultimate nature of reality. William James may have been referring to this in his discussions of "pure experience," which he characterized as "plain, unqualified actuality, or existence, a simple *that*," prior to the differentiations of subject and object and of mind and matter.[51] Buddhist contemplative insight is better characterized as becoming lucid during the waking state. In this truly "awakened" state of consciousness, we not only stop reifying phenomena, we directly perceive their empty nature, recognizing that all things exist only as dependently related events arising relative to a cognitive frame of reference.

While one is immersed in the nonconceptual realization of emptiness, all other phenomena vanish and one's awareness settles into ineffable bliss, luminosity, and spaciousness that far transcend analogous qualities of the substrate consciousness. Simply dwelling in the substrate consciousness after achieving quiescence is said to bring about no irreversible changes in one's view of reality, psychological well-being, or conduct. But the direct realization of emptiness allegedly has an enormous impact on all features of one's life, including the achievement of various paranormal abilities such as remote viewing and precognition.

The earlier discussion of a special theory and practices of ontological relativity mentioned paranormal abilities that may be achieved through the mental alteration of physical reality by manipulating archetypal forms associated with the form realm. Paranormal abilities stemming from the contemplative realization of emptiness do not rely on such manipulation. Rather, one can allegedly alter physical reality in much the same way that a lucid dreamer transforms events in a dream, by the power of the imagination and conceptual designation. Buddhists claim that there are definite limitations on the paranormal abilities of someone who has directly realized emptiness. No one, including a buddha, is omnipotent. The natural laws of karma place constraints on the degree to which one's own and others' realities may be altered by the power of the imagination. Nevertheless, there are numerous reports over the past two millennia of contemplatives who have achieved direct realization of emptiness displaying a wide range of paranormal abilities that were witnessed by other people.

If these claims are true—that such contemplatives can mentally manipulate the five elements in ways that they and others experience during the waking state, and can perceive events far removed in space and time—this

implies that their consciousness has transcended the substrate consciousness and even the form realm. This deeper dimension of awareness would
have to be nonlocal and atemporal, that is, it is not located in any particular
region of space or at any point in time. Michael Mensky's theories discussed in the preceding chapter may help explain the nature and potentials of this "superfluid" state of consciousness in terms of quantum physics. In his view, the most primal dimension of consciousness embraces
the whole quantum world, including all possible classical projections. People who have gained experiential access to that domain of consciousness
may deliberately emerge from the quantum world as a whole and look into
alternative realities other than that which is experienced in their immediate environment. At least from their own perspective, they might be able to
observe and thereby "make real" any of those alternatives.

Mensky likens a person who is trapped in the reified world of classical
physics to a horse wearing blinders so that it cannot look sideways.[52] Likewise, ordinary, "nonlucid" consciousness during the waking state is constrained by conceptual blinders, which place partitions between different
classical realities. Consequently, each component of such consciousness
apprehends only one reified world and makes decisions in accordance
with the information coming from that relatively stable and predictable
world. To apply Mensky's theory to Buddhist contemplative science, abiding in the direct, nonconceptual realization of emptiness, devoid of all
classical projections, is like being freed from all blinders. No classical reality has been selected, so one is not cognitively isolating one alternative reality from another. A facsimile of this occurs while in lucid, dreamless
sleep, experiencing only the substrate consciousness. When entering a lucid dream, a person may choose among a wide array of alternative dream
realities. But while dwelling in the nonconceptual realization of emptiness, one transcends the limitations of individual consciousness, space,
and time. And when emerging from this transpersonal state of consciousness and lucidly engaging with the intersubjective world, one may choose
from a much wider array of alternative waking realities.

Mensky points out that Everett's original interpretation of quantum mechanics, which predicts the same outputs of measurements as the Copenhagen interpretation, can be neither proved nor refuted in the framework
of physics. As long as experimental research is confined to the scientific
methods currently in use, which exclude contemplative training, it may be
impossible to either prove or disprove Everett's concept. Mensky comments, "This is a serious drawback, because constructing a rather (conceptually) complex interpretation that is impossible to verify seems to be too
high a price to be paid for making the theory more consistent in the purely

logical aspect."[53] His own theory, which he calls the Extended Everett's Concept, makes new predictions not found in usual quantum mechanics, but they are for features of consciousness rather than for the results of physical experiments. Therefore, according to Mensky, his theory can be tested using methods found in Buddhism for observing human consciousness.

As mentioned previously, Buddhism makes the astonishing claim that people who have directly realized emptiness may alter not only their own reality but also the realities of others. In certain circumstances, such contemplatives may manipulate the five elements in ways that can be witnessed by other people who do not share such insight. In other words, on occasion "miracles" may be demonstrated in ways that can be evaluated from a third-person perspective. But in Buddhism, these are not miracles in the sense of being supernatural events, any more than the discovery and amazing uses of lasers are miraculous—however they may appear to those ignorant of the nature and potentials of light. Such contemplatives claim to have realized the nature and potentials of consciousness far beyond anything known in contemporary science. What may appear supernatural to a scientist or a layperson may seem perfectly natural to an advanced contemplative, much as certain technological advances may appear miraculous to a contemplative.

Some historical perspective may be useful at this point. Contemporary scientific understanding of light is based on 400 years of research, during which physics has undergone two major revolutions. The scientific study of consciousness has been conducted, off and on, for little more than a century—with no revolution in the cognitive sciences—and there is still no scientific definition of consciousness or objective means of measuring it, and limited knowledge of its necessary and sufficient causes and its role in the natural world. According to the standards of any other branch of science, with respect to our understanding of consciousness, we are living in a dark age. But if we throw off our ethnocentric blinders, we may look into the alternative realities of other cultures and recognize that there is intelligent life outside the world of science as we know it. Although the above theories and alleged discoveries may be impossible to prove or disprove by means of physics alone, they may be put to the test of experience when scientific and contemplative methods of inquiry are integrated into a natural science that embraces the whole of nature.

## Evaluation

The theory of ontological relativity discussed above bears some similarity to the ancient Greek theme of Protagoras that "man is the measure of all things," implying that knowledge is invariably related to the knower.[54] In Plato's *Theaetetus*, this principle is applied to individual sensations of hot and cold. If one person sitting in a room declares that he feels hot and another person sitting right next to him says he feels cold, Protagoras maintained, it is impossible to contradict either one.[55] This indicates that judgments about qualities are inextricably related to subjective experience. His contemporaries were alarmed at the prospect of extending this principle to abstract qualities such as truth, beauty, justice, and virtue. Their concern was that when all objective criteria for good and evil were abandoned, the inevitable result would be moral relativism, which would lead to injustice and immorality.

Ironically, the modern scientific insistence on objectivity, entailing a marginalization of subjective experience, has resulted in just such moral relativism, which many people believe has led to widespread injustice and immorality. This methodological principle has a direct bearing on the scientific study of the mind, which is confined for the most part to third-person methods of inquiry: the collection of data from verbal reports of subjective experience, behavior, brain activity, and changes in hormone diffusion, heart rate, and other mentally related physiological processes. Contemporary cognitive scientists and philosophers defend this approach on the grounds that it accords with objective physical science and can (in principle) explain even the most private and ineffable subjective experiences, while never abandoning the third-person protocols of science.[56]

While this objectivist orientation has become the norm as a means to understand the nature and origins of mental phenomena, it is at variance with all other branches of the natural sciences. Imagine setting out to explore the quantitative world of mathematics by studying the verbal reports, brain states, and other behavioral and physiological measures of mathematicians. Or consider the prospects of understanding the qualities of various vintage wines by studying the verbal reports, behavior, brain states, and physiology of wine drinkers. You would certainly glean some superficial, inferential understanding of mathematics and wine, but for gaining thorough, empirical comprehension of these areas, these methods are ridiculously limited. Confining ourselves to such approaches for studying the mind and consciousness just doesn't make sense.

A major flaw in the objectivist strategy is the reliance upon subjects' beliefs about their own experiences, for in many cases, they perceive things that aren't there (such as visual illusions) and don't perceive things that are there (as in change blindness). Researchers attempt to identify such errors by cross-referencing subjective reports of experience with indirect, objective physiological and behavioral correlates of experience. Little if any progress has been made in terms of providing subjects with professional training so that they can become more accurate observers, experimenters, and reporters of their own experience. Unlike all other objects of scientific inquiry, mental phenomena have been left to observation by untrained amateurs.

The scientific study of the mind is thus limited to objective measurements in accordance with the principles of biology and anthropology. Researchers committed to this approach dismiss the hard problem of consciousness as if it doesn't exist or is of no significance, and block any real progress in solving it. In this way they are similar to quantum physicists who sidestep the measurement problem by glossing over the nature of observation, regarding their experimental systems of measurement from the perspective of classical physics. These two blind spots in the cognitive sciences and physics account for the continuing unresolved status of the hard problem and the measurement problem, which are likely to be closely interrelated.

Descartes viewed mind and matter as two independent, substantial classes of entities, a position that immediately creates the insurmountable problem of how the two could possibly interact. This is like looking at the worlds of subjective and objective events with double vision, never gaining depth perception of either one. Scientific naturalists have simplified this mind-matter problem by assuming that all mental phenomena are really (but inexplicably) physical phenomena in disguise. Consequently, most research is aimed at understanding states of consciousness from purely objective perspectives. With this elimination, or at least marginalization, of the first-person perspective, depth perception of physical realities remains impaired, and direct perception of mental realities is deliberately obstructed in a reductionist and impoverished view of the universe.

In its general theory of ontological relativity, the Middle Way view of Buddhism denies the independent existence of both mental and physical phenomena, and claims that all scientific, philosophical, aesthetic, and religious truths exist only relative to cognitive frames of reference. Critics have argued that this approaches extreme relativism, or even solipsism, in which not only subjective experiences but also issues of morality and the nature of the physical universe seem to depend on the arbitrary whims of the individual subject. For centuries, Buddhist philosophers have sought to avoid what they regard as the philosophical extremes of reification and nihilism. All Buddhist philosophical schools have grappled with the problem of how to draw the line between valid and invalid cognitions. Some adopt a pragmatic approach, arguing that the epistemological status of a cognition is determined by the desired or undesired outcome of an activity based on it. Others propose more normative criteria, arguing for a kind of correspondence between a subjective cognition and an objective state of affairs. In other words, valid cognitions have to be of something that "really" exists independently of anyone's awareness of it. Some Buddhist epistemologists conjoin these two criteria, suggesting that a cognition is valid if it both helps one to accomplish practical goals and correctly determines the nature of the object as it exists in its own right.[57]

Those representing the Middle Way view attempt to avoid the extremes of metaphysical realism and solipsism by proposing the following criteria for determining whether a hypothetical entity exists: it is apprehended relative to a cognitive frame of reference; its existence is not invalidated by more rigorously acquired empirical evidence or reasoning; and its existence is not repudiated by the ontological insight that nothing exists by its own absolute, inherent nature.[58] Without recourse to an assumed correspondence to anything that is real in and of itself, independent of any cognitive frame of reference, this is a kind of bootstrap method for investigating reality. Buddhist inquiry is in principle always evaluated in relation to the pursuit of genuine happiness (liberation and spiritual awakening) and virtue. So the above three epistemological criteria are closely aligned to pragmatic criteria as well.

Modern science has arguably progressed over the past four centuries precisely through these criteria: something is deemed to exist if it is observed and its existence is not repudiated by more rigorous means of observation, experiment, or analysis. Pragmatic criteria pertaining to the development of technology also play a major role in evaluating scientific knowledge. Generations of scientists have believed they were discovering, describing, and explaining objective realities as they exist in the real world, independent of any cognitive frame of reference, but this has been nothing more than a persistent and widespread illusion. The recognition of the relative nature of all scientific knowledge does nothing to invalidate it; it only reveals the error of Descartes' absolute division of reified subjects and objects, which is the metaphysical basis of classical physics and most of the rest of science.

The Middle Way view does allow for the possibility of errors even within the context of a single cognitive frame of reference. For example, one may introspectively mistake one's motivation for engaging in a certain activity. In terms of acquiring introspective knowledge, a single cognition can be evaluated—epistemically and pragmatically—only in relation to prior and subsequent cognitions. With training, one may enhance and refine the faculty of metacognition, or introspection, much as the human visual faculty has been enhanced and refined technologically with such instruments as the telescope. And through verbal discourse, one may cross-reference one's own experiences in relation to those of others who have engaged in such introspective training, much as scientists test each other's findings in their respective laboratories.

Errors in cognition may also occur across multiple cognitive frames of reference, and these may gradually be recognized in relation to earlier and later cognitions relative to one or more frames of reference. One common source of error is the assumption that a statement that is true for one cognitive frame of reference must be equally true for all other frames of reference. For example, if classical physics is adopted as background theory, one is bound to insist that quantum phenomena behave like classical phenomena, regardless of how they appear. What is true in classical physics must be true, one assumes, for all kinds of physics and other branches of science. However, if quantum physics is the background theory, the whole of classical physics and the rest of science are reassessed from that perspective.

In this chapter, I have been taking the Middle Way view as the background theory for evaluating all branches of knowledge, ancient and modern. Historically, religions have commonly posited their scriptures and other divine sources as their background theory, often assuming this to be an absolute frame of reference for determining all kinds of truths. But those claims have generally been eroded by other means of knowledge, especially science. Nowadays, many people have adopted science as the sole means of comprehending the natural world, but the evolution of science itself has empirically undermined any such idolization. Other disciplines of inquiry, such as the contemplative theories and practices of Buddhism, may eventually undermine the belief in science as the sole arbiter of knowledge of the universe. Perhaps some of the deepest insights of the contemplative traditions of the world have been at our fingertips for centuries, as expressed, for example, in these lines from Shakespeare:[59]

". . . These our actors
(As I foretold you) were all spirits, and
Are melted into air, into thin air,
And like the baseless fabric of this vision,
The cloud-capp'd tow'rs, the gorgeous palaces,
The solemn temples, the great globe itself,
Yea, all which it inherit, shall dissolve,
And like this insubstantial pageant faded
Leave not a rack behind. We are such stuff
As dreams are made on. . . ."

# 9

# PERFECT SYMMETRY

## A World Beyond Time

We return now to the hypothesis that quantum mechanics is universally correct, which inspired John Wheeler and Bryce DeWitt to adapt the Schrödinger equation as the wave function of the universe. A remarkable characteristic of this equation is that it portrays a universe that does not change with time; physicists call this the problem of frozen time, or simply the time problem. The gist is that the notion of evolution is not applicable to the universe as a whole, for it is assumed that there is no external observer with respect to the universe, and there is no external clock that does not belong to the universe.

Physicists may simply withdraw from this problem on the grounds that they are not actually trying to understand why the universe as a whole is evolving, they are just trying to understand their own experimental data. But this minimalist stance still begs the question of why we see the universe evolving in time in a given way. To try to solve this problem, we may fall back on a variation of Cartesian dualism, dividing the world into two domains: a subjective observer with his clock and other measuring devices, and the rest of the objective universe. But here quantum theory introduces an unexpected twist: the wave function of the rest of the universe depends on the designated time of the observer. This dependence is "objective" in the sense that the results obtained by different (macroscopic) observers living in the same quantum state of the universe and using a sufficiently good (macroscopic) measuring apparatus agree with each other.[1]

In plain language, the implication of this theory is that without introducing an observer, we have a dead universe that does not evolve in time, and this reemphasizes the role of the participant in the self-observing universe of quantum cosmology. The universe becomes alive (time-dependent) only when we think of it as divided into a subjective observer and the rest of the objective universe, and the wave function of the rest of the objective universe depends on the time measured by that observer. In other words, the evolution of the universe and everything in it, including life itself, is possible only with respect to the observer.[2]

The notion of an observer necessarily implies the presence of consciousness, without which no observation ever takes place, and the above theory implies that consciousness, far from being an insignificant by-product of brain activity, plays a crucial role in the formation and evolution of the universe. As Andrei Linde points out, the current scientific model of the material world obeying laws of physics has been so successful that we forget about our starting point—as conscious observers—and conclude that matter is the only reality and that perceptions are only helpful for describing it. But in fact, we are substituting the *reality* of our experience of the universe with a conceptually contrived *belief* in an independently existing material world.[3] The notion of time and a physically evolving universe independent of any observer is an illusion, albeit a very persistent one.[4]

Many physicists describe the world beyond time in terms of the melted vacuum, which embodies the laws of nature in their ideal, perfectly symmetrical state, while the frozen vacuum state of the universe in which we dwell reflects the current laws of nature. A fundamental difference between them is that the former is the lowest state of energy allowable by the laws of nature, while the latter is described as the lowest state of energy achievable by current technology. According to this view, over the course of cosmic evolution after the big bang, empty space gradually "froze," so that it has taken on internal structure like that of an ice crystal. From empty space emerged gravity, quarks, elementary particles, fields, and all other configurations of space-time and mass-energy. When space was still in its melted state, prior to the inflationary phase of the expanding universe, no such internal differentiations were present. Over the course of cosmic evolution, multiple symmetries were broken in ways that selected the universe we experience from among a wide array of alternative universes that might have emerged. This raises the questions: Did those symmetries have to break as they did, or might they have broken in other ways, resulting in other kinds of universes? And what caused our particular universe to evolve as it did? Might it have to do with our presence as observer-participants?[5]

The melted vacuum is an expression of perfect symmetry, a dimension of reality that has no internal structure and transcends time and change. If we could melt the current frozen structure of the universe, we would rediscover the perfect symmetry that existed before the universe cooled down. Physicists cannot directly observe this hidden perfection; they can only deduce its presence from clues provided by particle accelerators and highly energetic cosmic events. On the basis of such observations and mathematical analyses, the closer we trace the universe back to its origins, the closer we approach perfection, the most implicate of all orders of reality. The nature of that perfect vacuum may hold the key for understanding the universe as a whole. As Stanford physicist Leonard Susskind remarks, "Anybody who knows all about nothing knows everything."[6]

## The View of the Great Perfection

Many regard the pinnacle of Buddhist theory and practice to be the Great Perfection system of theory and practice resulting in perfect spiritual awakening. According to this view, the physical world, the form realm, and the formless realm all emerge from an implicate unity of the absolute space of phenomena (*dharmadhātu*), primordial consciousness (*jñāna*), and a primal energy (*jñāna-prāṇa*) that is indivisible from both space and consciousness. The absolute space of phenomena is not to be confused with relative space; rather, it is the ultimate dimension of reality out of which space, time, energy, matter, and mind all emerge. This primordial unity of space, consciousness, and energy is the ultimate implicate order.

Physicists have always set themselves the goal of understanding the objective universe as it exists independently of any relative observer, so their understanding of the melted and frozen vacuums is necessarily devoid of any notion of consciousness. This, as we have seen, may be a crucial limitation in their understanding of nature. Buddhists have always sought to understand the world of experience, not a purely objective world independent of experience. So in their understanding of nature, absolute space is not separate from primordial, nonlocal, time-transcending consciousness. And this ultimate consciousness is said to be imbued with unbounded knowledge and compassion and with a creative energy limited only by the natural laws of karma. This luminous space is the ground from which all possible worlds appear, and it is the ultimate nature of every observer's mind.

Much as physicists describe the current universe as "frozen" with respect to the perfect symmetry of the melted vacuum, so do Buddhists characterize our current minds as frozen with respect to the perfect symmetry of primordial consciousness. But that hidden perfection is not confined to the distant past, before our current "fall from grace." Rather, as the Dalai Lama comments, "Any given state of consciousness is permeated by the clear light of primordial awareness. However solid ice may be, it never loses its true nature, which is water. In the same way, even very obvious concepts are such that their 'place,' as it were, their final resting place, does not fall outside the expanse of primordial awareness. They arise within the expanse of primordial awareness and that is where they dissolve."[7] How is the perfect symmetry of this ultimate ground broken? In the words of Düdjom Lingpa, a nineteenth-century Tibetan master of the Great Perfection, "This ground is present in the mind-streams of all sentient beings, but it is tightly constricted by dualistic grasping; and it is regarded as external, firm, and solid. This is like water in its natural, fluid state freezing in a cold wind. It is due to dualistic grasping onto subjects and objects that the ground, which is naturally free, becomes frozen into the appearances of things."[8]

Like the melted vacuum of physics, the primordial unity of space, consciousness, and energy of the Great Perfection transcends time as we know it. Instead of being structured by the ordinary divisions of time, which are designated by specific observers within their own cognitive frames of reference, the Great Perfection is associated with "the fourth time," a dimension beyond the past, present, and future.[9] So the broken symmetries of relative space-time, mass-energy, and subject-object all emerge from the ultimate, undifferentiated symmetry of the absolute space of phenomena, the fourth time, primordial consciousness, and the energy of primordial consciousness, all of which are coextensive and of the same nature. These two sets of relative and ultimate phenomena have no inherent identities apart from the cognitive frameworks in which they are ascertained.

In this view, location in space-time is contingent upon the observer, but the emphasis is on the participant as a *perceiver,* not as a *conceptual designator.* Empirical observations exist only relative to the mode of perception and the technological system of measurement with which they are made. On a deeper level, theories exist only relative to the conceptual framework in which they are formulated. It is the participant as a *thinker* who establishes the demarcation between the measured system and the system of measurement and who establishes relative locality within space-time. This sets the universe—relative to a cognitive frame of reference—in motion. Without such participancy by a perceiving agent, there are no phenomena, and the universe is static. In other words, the multiple worlds of experience emerge into existence and evolve relative to the theory-laden experiences of observer-participants.

According to the cosmogony of the Great Perfection, all phenomena arise as displays of absolute space, which transcends all words and concepts, including the notions of existence and nonexistence, one and many, and subject and object. As a result of the delusional habit of reification, this infinite, luminous space is obscured and reduced to a blank, unthinking void, known as the substrate (*ālaya*). The experience of the substrate is like a dreamless sleep, devoid of appearances. From that void arises the substrate consciousness (*ālayavijñāna*), a state of limpid, clear consciousness from which all phenomena appear; it emerges from and is of the same nature as primordial consciousness. From the substrate consciousness arises the sense of self, or "I," which is apprehended as being "here," which results in the objective world appearing to be "over there," thus establishing the appearance of space. In this way, the dualistic experience of the world emerges from multiple, implicate orders of nonduality.

There are crucial differences between the substrate consciousness and primordial consciousness. When one's mind is settled in the substrate consciousness, one ascertains the nature of one's own mind in its relatively "frozen" state. Even though dualistic, discursive thoughts have subsided, this vacuum state of consciousness is subject to change and is implicitly structured by conceptual reification. The mind is temporarily in a state of relative equilibrium, or symmetry, but as soon as it emerges from that meditative state, the asymmetries of dualistic thinking are catalyzed as before. Primordial consciousness, in contrast, transcends time, and all appearances are present to it, without arising or ceasing. There is total knowledge and total awareness of all phenomena, without ever merging with or entering into objects. As Düdjom Lingpa explains, "Primordial consciousness is self-originating, naturally clear, free of outer and inner obscuration; it is the all-pervasive, radiant, clear infinity of space, free of contamination."[10]

Dualistic, or "frozen," consciousness is the natural radiance and clarity of the objects that emerge in the expanse of awareness. When they arise to our perceptual faculties, they are frozen by reification, as we grasp onto ourselves and all other things as inherently existent objects. The objective world is crystallized into separate and distinct things as a result of consciousness individually apprehending and labeling objects. They are experienced as agreeable, disagreeable, or neutral, and consequently thoughts of attachment to the agreeable, aversion to the disagreeable, and indifference to everything else emerge. Agreeable things are seen as good and become objects of hope, thus proliferating thoughts of yearning. Disagreeable things are seen as bad, and thus serve as a basis for thoughts of anxiety.

The way to return to the perfect symmetry of primordial consciousness is to realize how all phenomena fundamentally emerge from and are of the nature of absolute space. They have never existed except as displays of this primordial purity, so all appearances are illusory displays of our own primordial consciousness, which has taken on the guise of ordinary consciousness. It is not that consciousness must vanish into absolute space and primordial consciousness must arise from somewhere else. It just seems that way because of our ingrained tendency to reify ourselves and all objects of awareness.

In encountering the view of the Great Perfection, we first gain conceptual understanding based on verbal instruction, reading, study, and reflection. The next step is to investigate this theory, both analytically and experientially, until we fathom the lack of inherent existence of all objective and subjective phenomena. We now comprehend how they are all "empty" of any intrinsic identity, independent of any cognitive frame of reference. Finally, we comprehend how all things naturally, spontaneously arise from the expanse of the absolute space of phenomena and have no existence apart from that ultimate ground. We have now realized the view of the Great Perfection. To "gain confidence" in the view, we first identify the nature of primordial consciousness, then continually abide in that state of awareness until it remains unwaveringly at all times and in all situations.

While physicists speak of the perfect symmetry of the melted vacuum as a thing of the past, Buddhists regard the perfect symmetry of primordial consciousness as immanently present. According to Buddhist cosmogony, the form realm emerges from the formless realm, and the explicate order of the physical world emerges from the form realm. Eventually the reverse process will occur. But in every instant all three of these worlds spontaneously emerge from and dissolve back into the absolute space of phenomena. Just as the nature of ice is water, the nature of everything is the unity of primordial consciousness and absolute space. Once we cease objectifying ourselves and everything else and recognize the "one taste" of all phenomena as displays of primordial consciousness, we enter into a state of meditative equipoise in which all phenomena dissolve into the great expanse, with no object, obstruction, or intentionality.[11]

## The Way of the Great Perfection

To venture onto the path of the Great Perfection so that it leads to the perfect spiritual awakening of a buddha requires a great deal of theoretical and contemplative preparation. There is nothing to prevent people from trying to practice the Great Perfection with little or no background, but the results of a faint facsimile of authentic practice will be an equally faint facsimile of the results of authentic practice. One must be focused singlepointedly on this contemplative training, without being distracted by desires, concerns, and activities inconsistent with this discipline. In short, it is vital to turn away from all mundane concerns about material gain, transient pleasures, praise, and reputation, and to be content simply with merely adequate food, clothing, and lodging. It is very helpful in this regard to move away from one's habitual environment and circle of friends—in which ties of attachment are bound to be strong—and to devote oneself to solitary, highly focused practice day and night. Traditional manuals on this practice describe in much greater detail the specific kinds of preparatory meditations that support the practice of the Great Perfection.[12] In particular, if the training is to be fully effective, the prior accomplishment of meditative quiescence and insight into the empty nature of all phenomena are indispensable prerequisites.

The main practice consists of a thorough integration of the view, meditation, and way of life of the Great Perfection.[13] By means of sustained, rigorous study and critical analysis of this theory, supported by meditative quiescence and contemplative insight into emptiness, one may come to a profound certainty regarding the fundamental nature of one's own mind. To view one's mind from the perspective of the Great Perfection is to recognize that it has never been anything other than a display of primordial consciousness. When the mind is divested of all conceptual elaborations, including the constructs of subject and object and even existence and nonexistence, its essential nature is revealed as pristine awareness (Sanskrit: *vidyā*, Tibetan: *rig pa*). This is the primordial dimension of consciousness, which is neither contaminated by mental afflictions nor improved through spiritual practice. Abiding in the "fourth time," beyond the past, present, and future, it transcends all conceptual categories of arising and passing, permanence and impermanence, and existence and nonexistence. Its nature is primordially pure, empty, luminous, and all-pervasive; without internal differentiation, it is imbued with the perfection of all virtues.

Once we have acquired this view of the Great Perfection by "resting our awareness in its own state" and "seeing the true face of our own mind," the meditative practice consists simply of sustaining this awareness with unwavering, continuous mindfulness. Obviously, this is a far cry from simply dwelling in "choiceless awareness" or "just sitting," with no prior theoretical training, no basis in meditative quiescence, and no insight into emptiness. Without straying from this view, we release our awareness so that it is open to whatever phenomena appear to all our senses, without superimposing any thoughts or conceptual constructs onto them. Whatever thoughts occur of their own accord, we simply let them arise, without following after them or obstructing them. We attend to whatever arises with a sense of childlike wonder and freshness. When we sustain such awareness, without craving or aversion, all appearances—including all thoughts and emotions—arise as displays of primordial consciousness. If we view certain mental impulses as good or bad and reify them as such, the perfect symmetry of pristine awareness is broken, and our mind returns to its habitual frozen state. But as long as we sustain the awareness of all thoughts and other appearances as pure manifestations of primordial consciousness, then all the expressions of the mind spontaneously dissolve back into the open expanse of pristine awareness, without obscuring its true nature.

This meditative practice is to be sustained at all times, during all activities, day and night, without falling back to the habitual tendency of reifying either oneself or anything else.[14] The efficacy of such practice can be evaluated with respect to our dreams.[15] When we are well advanced in this training, our dreams are purified in the clear light of pristine awareness, and we become thoroughly immersed in this state of consciousness. Prior to that realization, we will be able to recognize our dreams and transform them at will. And in the early phases of this practice, we will at least find that negative dreams no longer arise.

On occasion, we will have experiences of bliss, luminosity, and nonconceptuality, and when that happens it is imperative to continue to rest in pristine awareness, with no craving for these qualities or aversion for their opposites. As insight deepens, unconditional love and compassion flow forth spontaneously, and we see beyond all distinctions of ultimate and relative truths. The one taste of all kinds of phenomena—from the most terrible to the most sublime—becomes apparent. With our mind transcending all personal preferences, even for spiritual awakening itself, our awareness rests in its own primordial ground, luminous and forever free.

For a way of life that supports and nurtures the view and meditation of the Great Perfection, it is vital to sustain an ongoing sense of compassion for all beings, without exception, and dedicate all the benefits of this practice to the welfare of others. According to Buddhist understanding, this training will catalyze mental imprints of previous negative behavior, or karma, and this results in various kinds of adversities by which the karma can be purified. These may include terrifying hallucinations either during the day or at night, attacks by other people, disease, and natural calamities. Mental afflictions may arise seemingly out of nowhere, and one may experience intense sadness and perplexity, as well as a wide array of other disturbing emotions. When such external or internal adversity strikes, it will actually enhance one's practice if one steadfastly abides in the view and the meditation. But if one reverts to viewing them with displeasure, fear, or aversion, this will derail the entire training. Even if we are well able to cope with adversity, we may have a harder time dealing with good fortune. If we respond to success, fame, and other mundane felicities with craving and attachment, this too will undermine our practice. Whatever happens, we do not blame anyone else for our troubles, but view everything from the perspective of the Great Perfection, with unwavering mindfulness.

## Spiritual Awakening

In the midst of a lucid dream, if we cease all activity and let even our thoughts subside, the entire dreamscape dissolves back into the substrate consciousness. If we continue to sustain lucid awareness, we will recognize the nature of our mind, free of appearances, in its relative ground state. In contrast, as the culmination of resting in primordial consciousness in the waking state, free of all activity, we enter the ultimate phase of practice of the Great Perfection, called "Extinction into Reality-Itself." All appearances dissolve into the innate unity of the absolute space of phenomena and primordial consciousness, and we realize the perfect spiritual awakening of a buddha. Our mind is forever free of all afflictions and obscurations and imbued with the perfection of all virtues, including wisdom and compassion. We continually experience a sense of blissful warmth and can live for months or even years on the food of *samādhi*, the power of bliss and emptiness.

According to traditional Buddhist accounts, it is possible for a buddha to manifest an inconceivable number of emanations in an unimaginable range of abodes of sentient beings, and in a single instant guide countless beings along the path of spiritual awakening. Extinction into Reality-Itself spontaneously results in mastery of the archetypal forms of the five elements, so one can transform one's body into any form of living being or inanimate object. Such emanations, it is said, are displayed in one's own and others' fields of experience; they are not simply subjective impressions. As one reengages with the world, everything appears as divine manifestations of the primordial ground.[16]

Although the realization of perfect awakening is the same for everyone, the external signs of this attainment vary from one person to the next. On rare occasions, recorded only a few times in the entire history of Tibet, adepts such as Padmasambhava are said to have displayed the "great transference rainbow body," in which their material body completely dissolves into the energy of primordial consciousness while they are still alive. Yet the appearance of their physical form remains and can be used at will, as if in a lucid dream. More commonly, the body of such an enlightened being dissolves at death into rainbow light, like a rainbow vanishes into space.

Penor Rinpoche, the former head of the Nyingma order of Tibetan Buddhism, recently claimed knowledge of six Tibetan contemplatives who manifested such a "rainbow body" during his lifetime. Such realization is revealed at death, when the contemplative's body gradually decreases in size until it vanishes altogether. In some cases, this dissolution process lasts as long as seven days, leaving only the hair and nails behind. This has been witnessed several times by many people over the past few decades.[17]

In a variation on Protagoras's assertion that "man is the measure of all things," the Buddha declared, "It is in this fathom-long body with its perceptions and its mind that I describe the world, the origin of the world, the cessation of the world, and the way leading to the cessation of the world."[18] The implication of this statement may be that if you thoroughly understand your body, in principle, you can fathom the nature of the physical universe. In this body you will find remnants of the big bang, all the elementary particles, and all the forces of nature—electromagnetic forces, strong forces, weak forces, and gravitational force—if you comprehend it thoroughly. But if you examine the physical organism alone, you will never fathom the mind. All the dimensions of consciousness and their relation to the objective world must be probed to their utter depths. This body-mind is therefore seen as the ideal laboratory for comprehending the entire universe, its origins, it cessation, and the path to spiritual awakening.

## Complementarities

One of the earliest references to the principle of complementarity is found in the Buddha's parable of the blind men and the elephant.[19] On one occasion, a number of his students commented to him that various scholars and philosophers engaged in seemingly endless debates about such topics as whether or not the universe is infinite and eternal and whether the soul dies with the body or lives on after death. In response, the Buddha told the parable of a king who called together a group of men who were born blind and brought an elephant into their presence. To one of them he presented the head of the elephant, to another its ears, to another a tusk, to others the trunk, foot, back, tail, and tuft of the tail. To each one he said, "Here is an elephant," and then asked them, one by one, to describe what they encountered. Depending on the part of the elephant that they had touched, they variously described the elephant as being like a pot, a winnowing basket, a ploughshare, a pillar, a pestle, and so on. When they heard one another's contradictory accounts, they immediately set to debating and quarreling about who was right, until eventually they came to blows. In the same way, the Buddha commented, people cling to their own views as being uniquely valid, then succumb to wrangling and even violence when others don't agree with them.

Returning to the previously discussed criteria—genuine happiness, understanding, and virtue—for evaluating theories and practices, Christianity has been deeply invested in all three from its earliest days. During the first four centuries of the Common Era, multiple schools of Christianity developed side by side. Among them, the Gnostic tradition emphasized belief, virtue, and knowledge as playing key roles on the path to salvation. The Roman Catholic Church, under the dominating influence of Augustine, placed a far greater emphasis on belief and faith in God as the source of all good; genuine happiness, beyond the world of change, was to be fully realized only in the hereafter. Only when the soul was united with its creator would "truth-given joy" be found. But the soul, according to Augustine, cannot be happy through any good of its own.[20]

Until the rise of modern science, Christians widely believed that their tradition represented the greatest hope for the realization of genuine happiness, understanding, and virtue. But from the seventeenth century onward, a growing number of Christian truth claims were discredited by advances in science, which evolved together with the Protestant movement in Christianity. In this new phase of the Christian Church, the gift of salvation had much more to do with simple faith and belief than with understanding or even virtue and good works. Within this theological context, Francis Bacon advocated his ideal of science as a means to understand nature in order to gain power over it and exploit it for human purposes. This goal, he was certain, was divinely sanctioned and to be accomplished with religious zeal. Descartes too predicted that by knowing the forces and the actions of material bodies, we can "make ourselves the masters and possessors of nature."[21] He also believed that the truths of mathematics are innate to the human mind, placed there by the hand of God. Galileo went further in regarding mathematics as the language of God, and this inspired him to seek above all a mathematical description of nature, as opposed to Descartes' emphasis on physical explanation.

A common theme among these Christian pioneers of the scientific revolution was the pursuit of a God's-eye view of reality, in which they envisioned knowing the mind of God through knowing his creation. The ultimate ideal of this religious and scientific quest was a kind of apotheosis, when man's understanding of the natural world merged with the understanding of God. With the unique human capacity of reason, these natural philosophers sought to conceptually understand nature as it truly exists, behind the veils of anthropocentric appearances to the physical senses.

Four hundred years later, modern science has made great progress in achieving the goals of its founders. Humanity has probed galactic clusters in the most distant regions of the universe, explored the nucleus of the atom, and probed the origins of the universe billions of years ago. This knowledge has brought us great power to dominate nature in ways never before imagined or accomplished in the history of the world. The world-view that has emerged is dominantly materialistic, with many scientists confidently asserting that the human mind and the rest of nature can finally be explained solely in terms of well-known physical processes. And many such scientists and philosophers are eager to dispute with or even suppress anyone who believes differently.[22] Science has made great contributions to our hedonic well-being. It has made great progress in medicine, curing, or at least managing, a wide range of physical and mental illnesses, and by way of technology it has provided a wide range of pleasures aroused by chemical, sensory, aesthetic, and intellectual stimuli. But it has contributed little, if anything, to the cultivation of genuine happiness as defined in this volume. Science has also yielded a tremendous amount of knowledge about the objective world of space-time and mass-energy, but its success in probing the mysteries of consciousness and the subjective world of mental phenomena has been far less impressive. In terms of the third criterion, virtue, science has offered little so far, either in understanding human virtues or in discovering methods for cultivating them. But the emergence of positive psychology has begun to fill that role.[23]

In our modern world, Christianity and science coexist, in tension with each other as they have been since the time of Copernicus. According to a recent Gallup poll, 83 percent of Americans and 49 percent of Europeans feel God is very important in their lives.[24] It is interesting to note that according to other recent surveys, 40 percent of the American scientists polled expressed a belief in a personal God to whom they can pray, which is roughly the same percentage as in a poll taken a century ago.[25] Such theists may formulate their own responses to the time problem mentioned at the beginning of this chapter. From an atheist perspective, the notion of evolution is not applicable to the universe as a whole, for there is no external observer with respect to the universe. But theists may counter that such an absolute, external observer does exist, and that, of course, is God. They may find this a reaffirmation of the biblical account of God creating the heavens and earth from nothing, a prominent theme of Christian and Jewish contemplatives since the early medieval era.[26]

Many people who are sympathetic both to science and to theism advocate a kind of complementarity between the two that allocates separate domains, or "nonoverlapping magisteria," to each.[27] Promoters of this version of conflict resolution present science and religion as independent and autonomous realms, each having its own domain of knowledge and methods to pursue its respective aims. The goal of science, in this view, is to explain the empirical realm of the objective universe with theories that are logically coherent and experimentally adequate, and to present quantitative predictions that can be tested experimentally. The goal of religion is to address questions concerning the meaning and purpose of life, our ultimate origins and destiny, and the experiences of our inner life. Religious texts, therefore, should not be read as scientific texts, and the claims of scientists should not be used to disprove the basis of religious belief. By confining themselves to their nonoverlapping domains and goals, science and religion should be able to coexist in a spirit of respectful noninterference.

The Buddhist tradition rejects both the materialism of modern science and the theological notion of a creator who exists independently of the universe and governs it, rewarding the virtuous and punishing the wicked. It also rejects the Cartesian dualism that underlies the above solution to the conflict between science and religion. Buddhism presents itself as an integrated system of theory and practice oriented toward the cultivation of genuine happiness, understanding, and virtue. Since the root of suffering is identified as ignorance and delusion, the primary means to liberation, or lasting, genuine happiness, must be valid insight into the nature of reality as a whole, including the entire world of experience. "The world in its variety arises from action," declares a classic fourth-century Buddhist text, presenting an observer-participancy view of reality in which worlds of experience emerge in relation to the acts of the sentient beings who inhabit them.[28] While science idealizes a conceptual, inferential understanding of the objective world as it exists independently of experience and Christianity idealizes faith in the truth of the word of God, Buddhism holds as its highest ideal direct, experiential insight into the nature of reality. This can be achieved only within the context of an ethical life, and yields genuine happiness while at the same time enhancing virtue.

There is a growing body of empirical evidence that Buddhist practices do in fact lead to greater happiness and virtue, but most Buddhist truth claims have yet to be put to the test of scientific inquiry. Whatever the merits of Buddhism may be in terms of understanding consciousness and its relation to reality as a whole, it has failed to produce vast knowledge of the natural sciences and has contributed nothing to technology.

In modern physics the theme of complementarity is closely associated with Niels Bohr, who declared that there are two kinds of truth, ordinary truth and deep truth. You can tell the difference between them, he said, by looking at their opposites, for the opposite of an ordinary truth is a falsehood, but the opposite of a deep truth is another deep truth. From this perspective, it may well be that science, Christianity, and Buddhism all embody deep, complementary truths and methods for achieving happiness, understanding, and virtue.

The complementarity of these views is closely related to the background views from which we interpret and evaluate empirical evidence and rational arguments. Committed atheists find that all truly scientific knowledge corroborates their naturalistic view of reality, and they are deeply skeptical, if not dismissive of anyone who believes otherwise. Committed theists often comment that scientific knowledge of creation constantly reaffirms their belief in the presence and active role of an all-knowing creator, and they are equally skeptical of those who fail to acknowledge this.[29] Some Buddhists similarly find that many of their beliefs have been corroborated by the latest advances in the natural sciences, and they express skepticism about the metaphysical claims of atheists and theists alike.[30] The notion that any one of these groups of believers is fundamentally more skeptical than the others is dubious at best. All are committed to their own cognitive frameworks, and they make sense of the world as they see it from their own perspectives.

While dogmatists of all varieties continue to battle among themselves, great advances in transportation and communication have brought humanity together in unprecedented ways. Some people retreat from this pluralistic, ever-changing world, while others embrace it as an extraordinary opportunity. We currently face a wide array of formidable problems that imperil our very existence on this planet. But we are also presented with a unique confluence of wisdom and practical insights from the world's civilizations. The need for us all to work together for the common good has never been greater, and the opportunities for doing so rise up before us with unprecedented splendor.

# NOTES

## 1. The Unnatural History of Science

1. John B. Watson, "Psychology as the Behaviorist Views It," *Psychological Review* XX (1913): 158–77; John B. Watson, *Behaviorism* (1913; reprint, New York: Norton, 1970); Arthur Koestler, *The Ghost in the Machine* (New York: Macmillan, 1967).

2. Alfred J. Ayer, *Language, Truth and Logic*, 2nd ed. (London: Gollancz, 1946), 90–94.

3. Paul M. Churchland, *Matter and Consciousness: A Contemporary Introduction to the Philosophy of Mind*, rev. ed. (Cambridge, MA: MIT Press, 1990); Stephen Stich, *From Folk Psychology to Cognitive Science: The Case Against Belief* (Cambridge, MA: Bradford, 1983).

4. John R. Searle, *Consciousness and Language* (Cambridge: Cambridge University Press, 2002), 9.

5. Owen Flanagan, *The Problem of the Soul: Two Visions of Mind and How to Reconcile Them* (New York: Basic Books, 2002), 88–94.

6. Francis Bacon, *Novum Organum* (1620), trans. and ed. P. Urbach and J. Gibson (Peru, IL: Open Court, 1994).

7. Eugene P. Wigner, "Remarks on the Mind-Body Question," in *Quantum Theory and Measurement*, ed. John Archibald Wheeler and Wojciech Hubert Zurek (Princeton: Princeton University Press, 1983), 178.

8. William James, *The Principles of Psychology* (1890; reprint, New York: Dover, 1950); *The Varieties of Religious Experience* (1902; reprint, Huntington, NY: Fontana, 1960).

9. Christof Koch, *The Quest for Consciousness: A Neurobiological Approach* (Englewood, CO: Roberts, 2004), 10.

10. Paul C.W. Davies, "An Overview of the Contributions of John Archibald Wheeler" in *Science and Ultimate Reality: Quantum Theory, Cosmology and Complexity, Honoring John Wheeler's 90th Birthday*, ed. John D. Barrow, Paul C.W. Davies, and Charles L. Harper Jr. (Cambridge: Cambridge University Press, 2004), 22.

11. Eugene Wigner, *Symmetries and Reflections* (Bloomington: Indiana University Press, 1967), 181.

## 2. The Many Worlds of Naturalism

1. Erwin Schrödinger, *The Interpretation of Quantum Mechanics* (Woodbridge, CT: Ox Bow Press, 1995).
2. P.C.W. Davies, "Particles Do Not Exist," in *Quantum Theory of Gravity*, ed. S. M. Christensen (Bristol: Adam Hilger, 1984); Michel Bitbol, *Schrödinger's Philosophy of Quantum Mechanics* (New York: Kluwer, 1995).
3. Roger Penrose, *Shadows of the Mind* (New York: Vintage, 1995), 419.
4. Richard Feynman, R. B. Leighton, and M. Sands, *The Feynman Lectures on Physics* (Reading, MA: Addison-Wesley, 1963), 4–2.
5. Henning Genz, *Nothingness: The Science of Empty Space*, trans. Karin Heusch (Cambridge, MA: Perseus, 1999), 26.
6. David Cook, *Probability and Schrödinger's Mechanics* (Hackensack, NJ: World Scientific, 2003), 6.
7. Jerome Bruner, *Beyond the Information Given: Studies in the Psychology of Knowing* (New York: Norton, 1973). For a review of neurobiological studies of such top-down influences, see Maurizio Corbetta and Gordon L. Shulman, "Control of Goal-Directed and Stimulus-Driven Attention in the Brain," *Nature Reviews Neuroscience* 3 (March 2002): 210–15.
8. Thomas Sprat, *The History of the Royal Society of London* (1667), ed. J. I. Cape and H. W. Jones (London: Routledge, 1959).
9. Lorraine Nicolas Remy, *Demonolatry* (1595), trans. E. A. Ashwin (London: University Books, 1930), xii.
10. Brian Easlea, *Witch-Hunting, Magic and the New Philosophy: An Introduction to Debates of the Scientific Revolution 1450–1750* (Brighton, NJ: Humanities Press, 1980).
11. For selected articles on M-Theory by Edward Witten, whom Brian Greene has called "Einstein's successor in the role of the world's greatest living physicist," see http://www.sns.ias.edu/~witten/.
12. Brian Greene, *The Elegant Universe: Superstrings, Hidden Dimensions, and the Quest for the Ultimate Theory* (New York: Norton, 1999).
13. Robert B. Laughlin, *A Different Universe: Reinventing Physics from the Bottom Down* (New York: Basic Books, 2005).
14. Albert Einstein, *Relativity: The Special and General Theory* (New York: Pi Press, 2005); Edwin Taylor and John A. Wheeler, *Space-Time Physics*, 2nd ed. (New York: W. H. Freeman, 1992).
15. M. B. Mensky, "Quantum Mechanics: New Experiments, New Applications, and New Formulations of Old Questions," *Physics—Uspekhi* 43, no. 6 (2000): 585–600.
16. Michael Lockwood, *Mind, Brain and the Quantum* (Oxford: Basil Blackwell, 1989), 20.
17. Michel Bitbol, "Materialism, Stances, and Open-Mindedness," in *Images of Empiricism: Essays on Science and Stances, with a Reply from Bas van Fraassen*, ed. Bradley Monton (Oxford: Oxford University Press, 2007).
18. Antonio Damasio, *The Feeling of What Happens: Body and Emotion in the Making of Consciousness* (New York: Harcourt, 1999), 322.

19. Thomas Metzinger, ed., *Neural Correlates of Consciousness: Empirical and Conceptual Questions* (Cambridge, MA: MIT Press, 2000).

20. Christof Koch, *The Quest for Consciousness: A Neurobiological Approach* (Englewood, CO: Roberts, 2004), 16–17.

21. Antonio Damasio, *Looking for Spinoza: Joy, Sorrow, and the Feeling Brain* (Orlando, FL: Harcourt, 2003); John R. Searle, *Mind: A Brief Introduction* (New York: Oxford University Press, 2004).

22. Owen Flanagan, *The Problem of the Soul: Two Visions of Mind and How to Reconcile Them* (New York: Basic Books, 2002).

23. David J. Chalmers, *Conscious Mind: In Search of a Fundamental Theory* (New York: Oxford University Press, 1996).

24. Eugene P. Wigner, "Remarks on the Mind-Body Question," in *Quantum Theory and Measurement*, ed. John Archibald Wheeler and Wojciech Hubert Zurek (Princeton: Princeton University Press, 1983), 175–79.

25. John R. Searle, *The Rediscovery of the Mind* (Cambridge, MA: MIT Press, 1994), 1.

26. John R. Searle, *Consciousness and Language* (Cambridge: Cambridge University Press, 2002), 49–50.

27. Erwin Schrödinger, *Nature and the Greeks* (New York: Columbia University Press, 1954), 6.

28. Koch, *The Quest for Consciousness*, 18–19.

## 3. Toward a Natural Theory of Human Consciousness

1. Stephen L. Adler, "Why Decoherence Has Not Solved the Measurement Problem: A Response to P. W. Anderson," *Studies in History and Philosophy of Science* 34 (2003): 135–42; Bernard d'Espagnat, *Veiled Reality: An Analysis of Present-Day Quantum Mechanical Concepts* (Reading, MA: Addison-Wesley, 1995), 177–89; Michael B. Mensky, *Quantum Measurements and Decoherence: Models and Phenomenology* (Dordrecht: Kluwer, 2000), 189–97.

2. Freeman J. Dyson, "Thought-Experiments in Honor of John Archibald Wheeler," in *Science and Ultimate Reality: Quantum Theory, Cosmology and Complexity, Honoring John Wheeler's 90th Birthday*, ed. John D. Barrow, Paul C.W. Davies, and Charles L. Harper Jr. (Cambridge: Cambridge University Press, 2004), 74.

3. B. Alan Wallace, *The Taboo of Subjectivity: Toward a New Science of Consciousness* (New York: Oxford University Press, 2000), 17–39.

4. Anton Zeilinger, "Why the Quantum? 'It' from 'Bit'? A Participatory Universe? Three Far-Reaching Challenges from John Archibald Wheeler and Their Relation to Experiment," in *Science and Ultimate Reality: Quantum Theory, Cosmology and Complexity, Honoring John Wheeler's 90th Birthday*, ed. John D. Barrow, Paul C.W. Davies, and Charles L. Harper Jr. (Cambridge: Cambridge University Press, 2004), 201.

5. Isaac Newton, *The Principia: Mathematical Principles of Natural Philosophy*, trans. I. Bernard Cohen and Anne Whitman (Berkeley: University of California Press, 1999).

6. J. W. Burrow, ed., *Charles Darwin: The Origin of Species* (London: Penguin, 1968).

7. Andrei Linde, "Inflation, Quantum Cosmology and the Anthropic Principle," in *Science and Ultimate Reality: Quantum Theory, Cosmology and Complexity, Honoring John Wheeler's 90th Birthday*, ed. John D. Barrow, Paul C.W. Davies, and Charles L. Harper Jr. (Cambridge: Cambridge University Press, 2004), 451.

8. Ibid., 454. Author's italics.

9. Michael B. Mensky, "Quantum Mechanics: New Experiments, New Applications, and New Formulations of Old Questions," *Physics—Uspekhi* 43, no. 6 (2000): 596. Author's italics.

10. Michael B. Mensky, "Concept of Consciousness in the Context of Quantum Mechanics," *Physics—Uspekhi* 48, no. 4 (2005): 390.

11. Vitaly L. Ginzburg, *About Science, Myself, and Others* (Bristol: Institute of Physics Publications, 2005), 54.

12. Mensky, "Concept of Consciousness in the Context of Quantum Mechanics," 390.

13. M. Jibu and K. Yasue, *Quantum Brain Dynamics and Consciousness—An Introduction* (Amsterdam: John Benjamins, 1995).

14. C. P. Enz, "On Preparata's Theory of a Super Radiant Phase Transition," *Helvetica Physica Acta* 70 (1997): 141–53.

15. Paul C.W. Davies, "An Overview of the Contributions of John Archibald Wheeler," in *Science and Ultimate Reality: Quantum Theory, Cosmology and Complexity, Honoring John Wheeler's 90th Birthday*, ed. John D. Barrow, Paul C.W. Davies, and Charles L. Harper Jr. (Cambridge: Cambridge University Press, 2004), 6.

16. Ibid.

17. Ibid., 3.

## 4. Observing the Space of the Mind

1. Immanuel Kant, *Metaphysical Foundations of Natural Science*, trans. James Ellington. (Indianapolis: Bobbs Merrill, 1970), preface, AK IV, 471.

2. Michael B. Mensky, "Quantum Mechanics: New Experiments, New Applications, and New Formulations of Old Questions," *Physics—Uspekhi* 43, no. 6 (2000): 396.

3. B. Alan Wallace, *The Taboo of Subjectivity: Toward a New Science of Consciousness* (New York: Oxford University Press, 2000), 59–73.

4. Edmund Husserl, *Ideas Pertaining to a Pure Phenomenology and to a Phenomenological Philosophy*, trans. Ted E. Klein and William E. Pohl (Boston: M. Nijhoff, 1980), vol. 1.

5. Ludwig Wittgenstein, *Philosophical Investigations*, trans. G.E.M. Anscombe (Oxford: Blackwell, 1958), sections 244–271.

6. Peter Machamer, ed., *The Cambridge Companion to Galileo* (Cambridge: Cambridge University Press, 1998), 64f.

7. Sigmund Freud, *New Introductory Lectures on Psychoanalysis*, trans. and ed. James Strachey (New York: Norton, 1989).

8. L. Postman, J. Bruner, and R. Walk, "The Perception of Error," *British Journal of Psychology* 42 (1951): 1–10.

9. Gilbert Ryle, *The Concept of Mind* (London: Hutchinson, 1963), ch. 1.

10. On inwardness, introspection, and the modern study of consciousness, see B. F. Skinner, *Science and Human Behavior* (New York: Macmillan, 1953); William Lyons, *The Disappearance of Introspection* (Cambridge, MA: MIT Press, 1986); Charles Taylor, *Sources of the Self: The Making of the Modern Identity* (Cambridge, MA: Harvard University Press, 1989); Sydney Shoemaker, *The First-Person Perspective and Other Essays* (Cambridge: Cambridge University Press, 1996); Wallace, *The Taboo of Subjectivity*, 75–120; B. Alan Wallace, *Balancing the Mind: A Tibetan Buddhist Approach to Refining Attention* (Ithaca, NY: Snow Lion, 2005), 269–96; B. Alan Wallace, *Contemplative Science: Where Buddhism and Neuroscience Converge* (New York: Columbia University Press, 2006).

11. B. Alan Wallace, *Genuine Happiness: Meditation as the Path to Fulfillment* (Hoboken, NJ: Wiley, 2005), 11–21; B. Alan Wallace, *The Attention Revolution: Unlocking the Power of the Focused Mind* (Boston: Wisdom, 2006), 13–68.

12. Wallace, *Genuine Happiness*, 22–34; Wallace, *The Attention Revolution*, 77–124.

13. Wallace, *The Attention Revolution*, 49–55.

14. For a Buddhist view of objectivity, see Wallace, *Contemplative Science*, ch. 4.

15. For information on the Shamatha Project, see http://sbinstitute.com and http://mindbrain.ucdavis.edu/index_html.

16. Wallace, *Genuine Happiness*, 65–90.

17. Anne C. Klein, "Mental Concentration and the Unconditioned: A Buddhist Case for Unmediated Experience," in *Paths to Liberation: The Mārga and Its Transformations in Buddhist Thought*, ed. Robert E. Buswell Jr. and Robert M. Gimello (Honolulu: University of Hawaii Press, Studies in East Asian Buddhism 7, 1992), 269–308; Robert K. C. Forman, ed., *The Problem of Pure Consciousness: Mysticism and Philosophy* (New York: Oxford University Press, 1990); Jonathan Shear, *The Inner Dimension: Philosophy and the Experience of Consciousness* (New York: P. Lang, 1990); Robert K.C. Forman, ed., *The Innate Capacity: Mysticism, Psychology, and Philosophy* (New York: Oxford University Press, 1998); Robert K.C. Forman, *Mysticism, Mind, Consciousness* (Albany: State University of New York Press, 1999).

18. Wallace, *The Attention Revolution*, 155–65.

19. Wallace, *Genuine Happiness*, 183–95.

20. B. Alan Wallace, "Vacuum States of Consciousness: A Tibetan Buddhist View," in *Buddhist Thought and Applied Psychology: Transcending the Boundaries*, ed. D. K. Nauriyal (London: Routledge-Curzon, 2006), 112–21.

21. Düdjom Lingpa, *The Vajra Essence: From the Matrix of Pure Appearances and Primordial Consciousness, a Tantra on the Self-Originating Nature of Existence*, trans. B. Alan Wallace (Alameda, CA: Mirror of Wisdom, 2004), 92.

22. For one of the earliest accounts of the Buddha's alleged recollections of his own and others' past lives, see *Majjhima Nikāya*, 36.38–40 in *The Middle Length Discourse of the Buddha*, trans. Bhikkhu Ñāṇamoli and Bhikkhu Bodhi (Boston: Wisdom, 1995), 341. For instructions on how to develop recall of past lives once one has achieved a high level of meditative concentration, see Buddhaghosa, *The Path of Purification*, trans. Ñāṇamoli Bhikkhu (Kandy: Buddhist Publication So-

ciety, 1979), XIII:13–120; Geshe Gedün Lodrö, *Walking Through Walls: A Presentation of Tibetan Meditation*, trans. and ed. Jeffrey Hopkins (Ithaca, NY: Snow Lion, 1992), 287–88.

23. Richard P. Feynman, *The Character of Physical Law* (Cambridge, MA: MIT Press, 1983), 158.

24. R. Waterfield, ed., *The First Philosophers: The Presocratics and Sophists Translated with Commentary* (Oxford: Oxford University Press, 2000), F9.

## 5. A Special Theory of Ontological Relativity

1. John R. Searle, *Consciousness and Language* (Cambridge: Cambridge University Press, 2002), 34.

2. John R. Searle, *Mind: A Brief Introduction* (New York: Oxford University Press, 2004), 268.

3. L. Postman, J. Bruner, and R. Walk, "The Perception of Error," *British Journal of Psychology* 42 (1951): 1–10; Jerome Bruner, *Beyond the Information Given: Studies in the Psychology of Knowing* (New York: Norton, 1973).

4. Plato, *Timaeus*, trans. Benjamin Jowett, (Indianapolis: Bobbs-Merrill, 1949), 53c–55c, Nom. 886d.

5. Werner Heisenberg, *Physics and Philosophy* (London: Penguin, 1989), 59; author's italics; see also E. Schrödinger, *The Interpretation of Quantum Mechanics* (Woodbridge, CT: Ox Bow Press, 1995); H. D. Zeh, "There Are no Quantum Jumps, nor Are There Particles," *Physics Letters* A172 (1993): 189–92; P.C.W. Davies, "Particles Do Not Exist," in *Quantum Theory of Gravity*, ed. S. M. Christensen (New York: Adam Hilger, 1984); Michel Bitbol, *Schrödinger's Philosophy of Quantum Mechanics* (New York: Kluwer, 1995).

6. Werner Heisenberg, *Daedalus: Journal of the American Academy of Arts and Sciences* 87 (1958): 95.

7. John D. Barrow, "Outward Bound," in *Spiritual Information: 100 Perspectives on Science and Religion*, ed. Charles L. Harper Jr. (West Conshohocken, PA: Templeton Foundation Press, 2005), 118.

8. Harald Atmanspacher and Hans Primas, "Pauli's Ideas on Mind and Matter in the Context of Contemporary Science," *Journal of Consciousness Studies* (in press).

9. Harald Atmanspacher and Hans Primas, "The Hidden Side of Wolfgang Pauli," *Journal of Consciousness Studies* 3 (1996): 112–26.

10. C. G. Jung, *The Collected Works of C. G. Jung. Volume 14. Mysterium Conjunctionis*, 2nd ed. (Princeton: Princeton University Press, 1970), par. 767.

11. Letter by Pauli to Fierz, August 12, 1948. Letter 971 in K. von Meyenn, ed., *Wolfgang Pauli. Wissenschaftlicher Briefwechsel, Band III: 1940–1949*, trans. Harald Atmanspacher and Hans Primas (Berlin: Springer, 1993), 559.

12. David Bohm, "A New Theory of the Relationship of Mind and Matter," *Philosophical Psychology* 3 (1990): 271–86. See also Harald Atmanspacher and Fred Kronz, "Relative Onticity," in *On Quanta, Mind, and Matter*, ed. H. Atmanspacher, A. Amann, and U. Müller-Herold (Dordrecht: Kluwer, 1999), 273–94; Harald

Atmanspacher, "Mind and Matter as Asymptotically Disjoint, Inequivalent Representations with Broken Time-Reversal Symmetry," *BioSystems* 68 (2003): 19–30.

13. Eugene P. Wigner, "Physics and the Explanation of Life," *Foundations of Physics* 1 (1970): 35–45; Bernard d'Espagnat, "Aiming at Describing Empirical Reality," in *Potentiality, Entanglement, and Passion-at-a-Distance*, ed. R. S. Cohen, M. Horne, and J. Stachel (Dordrecht: Kluwer, 1997), 71–87; Bernard d'Espagnat, "Concepts of Reality," in *On Quanta, Mind, and Matter*, ed. H. Atmanspacher, A. Amann, and U. Müller-Herold (Dordrecht: Kluwer, 1999), 249–70; N. D. Mermin, "What Is Quantum Mechanics Trying to Tell Us?" *American Journal of Physics* 66 (1998): 753–67.

14. Jacob D. Bekenstein, "Information in the Holographic Universe," *Scientific American* 9, no. 2 (August 2003).

15. Leonard Susskind, "The World as a Hologram," *Journal of Mathematical Physics* 36 (1995): 6377–6396; Leonard Susskind, "Black Holes and the Information Paradox," *Scientific American* 276, no. 4 (April 1997): 52–57; Raphael Bousso, "The Holographic Principle," *Reviews of Modern Physics* 74 (2002): 825–74.

16. George F. R. Ellis, "True Complexity and Its Associated Ontology," in *Science and Ultimate Reality: Quantum Theory, Cosmology and Complexity, Honoring John Wheeler's 90th Birthday*, ed. John D. Barrow, Paul C.W. Davies, and Charles L. Harper Jr. (Cambridge: Cambridge University Press, 2004), 607–36.

17. George F. R. Ellis, "Progress in Scientific and Spiritual Understanding," in *Spiritual Information: 100 Perspectives on Science and Religion*, ed. Charles L. Harper Jr. (West Conshohocken, PA: Templeton Foundation Press, 2005), 130.

18. Ellis, "True Complexity and Its Associated Ontology," 631.

19. Roger Penrose, *The Emperor's New Mind* (Oxford: Oxford University Press, 1989).

## 6. High-Energy Experiments in Consciousness

1. http://www.absoluteastronomy.com/reference/large_hadron_collider.

2. Pa-Auk Tawya Sayadaw, *Knowing and Seeing* (Kuala Lumpur, Malaysia: WAVE Publications, 2003), 51–53.

3. "What is the starting of wholesome states? Virtue that is well purified and view that is straight." *Saṃyutta Nikāya*, V:143 in *The Connected Discourses of the Buddha*, trans. Bhikkhu Bodhi (Boston: Wisdom, 2000), 1629.

4. *Dantabhūmi Sutta (Majjhima Nikāya, 125)* in *The Middle-Length Discourse of the Buddha*, trans. Bhikkhu Ñāṇamoli and Bhikkhu Bodhi (Boston: Wisdom, 1995), 989–96.

5. Tsong-kha-pa, *The Great Treatise on the Stages of the Path to Enlightenment*, trans. The Lamrim Chenmo Translation Committee (Ithaca, NY: Snow Lion, 2002), 3:28–30.

6. Buddhaghosa, *The Path of Purification*, trans. Ñāṇamoli Bhikkhu (Kandy: Buddhist Publication Society, 1979), IV.

7. Sayadaw, *Knowing and Seeing*, 55.

8. B. Alan Wallace, *The Attention Revolution: Unlocking the Power of the Focused Mind* (Boston: Wisdom, 2006), chapter 10.

9. Buddhaghosa, *The Path of Purification*, V:1–42; Vasubandhu, *Abhidharmakośabhāṣ yam*, trans. Louis de La Vallée Poussin; English trans. Leo M. Pruden (Berkeley: Asian Humanities Press, 1991), VIII:36.

10. Buddhaghosa, *The Path of Purification*, III:74–103, 121–22.

11. Sayadaw, *Knowing and Seeing*, 77–79.

12. Paravahera Vajirañāṇa, *Buddhist Meditation in Theory and Practice* (Kuala Lumpur, Malaysia: Buddhist Missionary Society, 1975), 145.

13. www.paauk.org.

14. Vajirañāṇa, *Buddhist Meditation in Theory and Practice*, 37.

15. Buddhaghosa, *The Path of Purification*, I:43.

16. *Dīgha Nikāya*, I:73 in *The Long Discourse of the Buddha*, trans. Maurice Walsh (Boston: Wisdom, 1995), 102; *Aṅguttara Nikāya*, III:21; Kamalaśīla, *First Bhāvanākrama*, in *Minor Buddhist Texts, Part II*, ed. G. Tucci (Rome: Istituto italiano per il Medio ed Estremo Oriente 1958), 205.

17. Vasubandhu, *Abhidharmakośabhāṣyam*, III.

18. Vasubandhu, *Abhidharmakośabhāṣyam*, VIII.

19. Jamgön Kongtrul Lodrö Tayé, *Myriad Worlds: Buddhist Cosmology in Abhidharma, Kālacakra and Dzog-chen*, trans. International Translation Committee of Kunkhyab Chöling (Ithaca, NY: Snow Lion, 1995), 155.

20. Buddhaghosa, *The Path of Purification*, XII–XIII.

21. Hammalawa Saddhatissa, *Buddhist Ethics* (Boston: Wisdom, 1997), 143.

22. Jamgön Kongtrul Lodrö Tayé, *Buddhist Ethics*, trans. The International Translation Committee (Ithaca, NY: Snow Lion, 1998).

23. Michael Riordan, "Science Fashions and Scientific Fact," http://www.physics today.org/vol-56/iss-8/p50.html#bio.

24. B. Alan Wallace, *Contemplative Science: Where Buddhism and Neuroscience Converge* (New York: Columbia University Press, 2006), chapter 4.

25. Ernan McMullin, "Enlarging the Known World," in *Physics and Our View of the World*, ed. Jan Hilgevoord (Cambridge: Cambridge University Press, 1994), 79–113.

26. Francis Bacon, *Novum Organum*, trans. and ed. P. Urbach and J. Gibson (Peru, IL: Open Court, 1994).

27. Klaus Michael Meyer-Abich, "Science and Its Relation to Nature in C. F. von Weizsäcker's "Natural Philosophy," in *Time, Quantum and Information*, ed. Lutz Castell and Otfried Ischebeck (Berlin: Springer Verlag, 2003), 173–85.

28. Carl Friedrich von Weizsäcker, *The History of Nature* (London: Routledge & Kegan Paul, 1951), 179.

## 7. A General Theory of Ontological Relativity

1. Ludwig Wittgenstein, *On Certainty*, trans. Denis Paul and G.E.M. Anscombe (San Francisco: Arion Press, 1991), 83, 306.

2. Willard Van Orman Quine, *Ontological Relativity and Other Essays* (New York: Columbia University Press, 1969), 51–64; Willard Van Orman Quine, *From Stimulus to Science* (Cambridge, MA: Harvard University Press, 1995).

3. Bas Van Fraassen, "From Vicious Circle to Infinite Regress and Back Again," in *Proceedings of the 1992 Biennial Meeting of the Philosophy of Science Association*, vol. 2, ed. D. Hull, M. Forbes, and K. Okruhlick (Chicago: University of Chicago Press, 1993).

4. Bas Van Fraassen, *The Scientific Image* (New York: Oxford University Press, 1980), 58.

5. Bas Van Fraassen, *The Empirical Stance* (New Haven: Yale University Press, 2002), 139.

6. Harvard philosopher Hilary Putnam has developed a theory of "pragmatic realism" that has much in common with the theory presented here. See Hilary Putnam, *The Many Faces of Realism* (La Salle, IL: Open Court, 1987); Hilary Putnam, *Realism with a Human Face*, ed. James Conant (Cambridge, MA: Harvard University Press, 1990); and B. Alan Wallace, *The Taboo of Subjectivity: Toward a New Science of Consciousness* (New York: Oxford University Press, 2000), 63–67.

7. Paul C.W. Davies, "An Overview of the Contributions of John Archibald Wheeler," in *Science and Ultimate Reality: Quantum Theory, Cosmology and Complexity, Honoring John Wheeler's 90th Birthday*, ed. John D. Barrow, Paul C.W. Davies, and Charles L. Harper Jr. (Cambridge: Cambridge University Press, 2004), 10.

8. Carl Friedrich von Weizsäcker, *The Unity of Nature*, trans. Francis J. Zucker (New York: Farrar, Straus & Giroux, 1980), 406.

9. Recall Einstein's often-quoted statement, "I believe in Spinoza's God who reveals himself in the orderly harmony of what exists, not in a God who concerns himself with fates and actions of human beings." *New York Times*, April 25, 1929, p. 60, col. 4.

10. John Wild, ed., *Spinoza: Selections* (New York: Charles Scribner's Sons, 1930).

11. Albert Einstein, *Ideas and Opinions* (New York: Crown, 1954), 262.

12. Časlav Brukner and Anton Zeilinger, "Information and Fundamental Elements of the Structure of Quantum Theory," in *Time, Quantum and Information*, ed. Lutz Castell and Otfried Ischebeck (Berlin: Springer Verlag, 2003), 323–55.

13. Anton Zeilinger, "Why the Quantum? 'It' from 'Bit'? A Participatory Universe? Three Far-Reaching Challenges from John Archibald Wheeler and Their Relation to Experiment," in *Science and Ultimate Reality: Quantum Theory, Cosmology and Complexity, Honoring John Wheeler's 90th Birthday*, ed. John D. Barrow, Paul C.W. Davies, and Charles L. Harper Jr. (Cambridge: Cambridge University Press, 2004), 201–20; Anton Zeilinger, "The Message of the Quantum," *Nature* 438, no. 8 (December 2005).

14. Albert Einstein, "Autobiographical Notes," in ref. [27] quotation, in P. A. Schlipp, *Albert Einstein: Philosopher-Scientist* (Evanston, IL: Library of Living Philosophers, 1949), 81.

15. Niels Bohr, *Essays 1958–1962 on Atomic Physics and Human Knowledge* (New York: Wiley, 1963), 3.

16. Niels Bohr, *Atomic Theory and the Description of Nature* (Cambridge: Cambridge University Press, 1934), 18.

17. Martin Curd and J. A. Cover, eds., "The Duhem-Quine Thesis and Underdetermination," in *Philosophy of Science* (New York: Norton, 1998), Section 3; W. V. Quine, "On Empirically Equivalent Systems of the World." *Erkenntnis* 9 (1975): 313–28.

18. Freeman J. Dyson, "Thought-Experiments in Honor of John Archibald Wheeler," in *Science and Ultimate Reality: Quantum Theory, Cosmology and Complexity, Honoring John Wheeler's 90th Birthday,* ed. John D. Barrow, Paul C.W. Davies, and Charles L. Harper Jr. (Cambridge: Cambridge University Press, 2004), 84.

19. John Archibald Wheeler, "Law Without Law," in *Science and Ultimate Reality: Quantum Theory, Cosmology and Complexity, Honoring John Wheeler's 90th Birthday,* ed. John D. Barrow, Paul C.W. Davies, and Charles L. Harper Jr. (Cambridge: Cambridge University Press, 2004), 184.

20. Albert Einstein, *Relativity: The Special and the General Theory* (New York: Three Rivers Press, 1961), vii, 170; A. Forsee, *Albert Einstein, Theoretical Physicist* (New York: Macmillan, 1963), 81.

21. George Greenstein and Arthur G. Zajonc, *The Quantum Challenge: Modern Research on the Foundations of Quantum Mechanics* (Boston: Jones and Bartlett, 1997), 38–42.

22. Davies, "An Overview of the Contributions of John Archibald Wheeler," 9.

23. Wheeler, "Law Without Law," 209.

24. E. R. Harrison, *Cosmology: The Scientific Universe* (New York: Cambridge University Press, 1981), 2; George Greenstein, *The Symbiotic Universe: Life and Mind in the Cosmos* (New York: Morrow, 1988); John D. Barrow and Frank J. Tipler, *The Anthropic Cosmological Principle* (Oxford: Oxford University Press, 1996).

25. B. Alan Wallace, *Choosing Reality: A Buddhist View of Physics and the Mind* (Ithaca, NY: Snow Lion, 1996), 108–12.

26. S. Langer, *Philosophy in a New Key. A Study in the Symbolism of Reason, Rite, and Art,* 3rd ed. (Cambridge, MA: Harvard University Press, 1978), 273.

27. Wheeler, "Law Without Law," 196; Zeilinger, "Why the Quantum?," 209.

28. David Finkelstein, "Ur Theory and Space-time Structure," in *Time, Quantum and Information,* ed. Lutz Castell and Otfried Ischebeck (Berlin: Springer Verlag, 2003), 400.

29. Hugh Everett, "Short Article," *Reviews of Modern Physics* 29 (1957): 454.

30. B. S. DeWitt and N. Graham, eds., *The Many-Worlds Interpretation of Quantum Mechanics* (Princeton: Princeton University Press, 1973).

31. Michael B. Mensky, "Quantum Mechanics: New Experiments, New Applications, and New Formulations of Old Questions," *Physics—Uspekhi* 43, no. 6 (2000): 585–600; "Concept of Consciousness in the Context of Quantum Mechanics," *Physics—Uspekhi* 48, no. 4 (2005): 389–409.

32. Mensky, "Concept of Consciousness in the Context of Quantum Mechanics," 399.

33. Ibid., 405.

34. Ibid., 408.

## 8. Experiments in Quantum Consciousness

1. Arthur Zajonc, ed., *The New Physics and Cosmology: Dialogues with the Dalai Lama* (New York: Oxford University Press, 2004).
2. http://www.quantum.univie.ac.at/links/sci_am/teleportation.pdf.
3. Anton Zeilinger's homepage at the University of Vienna is http://www.quantum.univie.ac.at/zeilinger/.
4. Erwin Schrödinger, *Mind and Matter* (Cambridge: Cambridge University Press, 1958), 38.
5. Vasubandhu, *Abhidharmakośabhāṣyam*, French trans. Louis de La Vallée Poussin; English trans. Leo M. Pruden (Berkeley: Asian Humanities Press, 1991), III:85, fn. 483; Jamgön Kongtrul Lodrö Tayé, *Myriad Worlds: Buddhist Cosmology in Abhidharma, Kālacakra and Dzog-chen*, trans. and ed. the International Translation Committee (Ithaca, NY: Snow Lion, 1995), 167–69.
6. Zajonc, ed., *The New Physics and Cosmology*, 91.
7. David Ritz Finkelstein, "Emptiness and Relativity," in *Buddhism and Science: Breaking New Ground*, ed. B. Alan Wallace (New York: Columbia University Press, 2003), 365–84.
8. Nārada Mahā Thera, *A Manual of Abhidhamma*, 4th rev. ed. (Kuala Lumpur, Malaysia: Buddhist Missionary Society, 1979), 109, 216.
9. Vasubandhu, *Abhidharmakośabhāṣyam*, III:85, fn. 484.
10. Tayé, *Myriad Worlds*, 167–69.
11. B. Alan Wallace, *Balancing the Mind: A Tibetan Buddhist Approach to Refining Attention* (Ithaca, NY: Snow Lion, 2005).
12. Tsong-kha-pa, *The Great Treatise on the Stages of the Path to Enlightenment* (Ithaca, NY: Snow Lion, 2002), 3:92, 96–97.
13. B. Alan Wallace, *The Attention Revolution: Unlocking the Power of the Focused Mind* (Boston: Wisdom, 2006), chapter 10.
14. Ibid., 90.
15. *Majjhima Nikāya*, in *The Middle-Length Discourse of the Buddha*, trans. Bhikkhu Ñāṇamoli and Bhikkhu Bodhi (Boston: Wisdom, 1995), 36, 85, 100.
16. Tsong-kha-pa, *The Great Treatise on the Stages of the Path to Enlightenment*, 3:96–98; Padmasambhava, *Natural Liberation: Padmasambhava's Teachings on the Six Bardos*, comm. Gyatrul Rinpoche; trans. B. Alan Wallace (Boston: Wisdom, 1998), 113–15; H. H. the Dalai Lama, Dzong-ka-ba, and Jeffrey Hopkins, *Yoga Tantra: Paths to Magical Feats* (Ithaca, NY: Snow Lion, 2005), 104–108.
17. D. M. Strong, trans., *The Udāna, or the Solemn Utterances of the Buddha* (Oxford: Pali Text Society, 1994), I:10; *Māluṅkyāputta Sutta, Sutta Nipāta*, in *The Middle-Length Discourse of the Buddha*, trans. Bhikkhu Ñāṇamoli and Bhikkhu Bodhi (Boston: Wisdom, 1995), XXXV:95.
18. *Milindapañha*, 37–38; Buddhaghosa, *The Path of Purification*, trans. Ñāṇamoli Bhikkhu (Kandy: Buddhist Publication Society, 1979), XIV:141; R. M. L. Gethin, *The Buddhist Path to Awakening* (Oxford: Oneworld Publications, 2001), 36–44; Ven. Anālayo, "Mindfulness in the Pāli *Nikāyas*," in *Buddhist Thought and Ap-*

*plied Psychology: Transcending the Boundaries*, ed. D. K. Nauriyal (London: Routledge-Curzon, 2006), 229–49.

19. *Satipaṭṭhāna Sutta*, in *The Middle-Length Discourses of the Buddha*, trans. Bhikkhu Bodhi (Boston: Wisdom, 1995), 145–55.

20. *Satipaṭṭhānasutta* 5, 146.

21. B. Alan Wallace, *Genuine Happiness: Meditation as the Path to Fulfillment* (Hoboken, NJ: Wiley, 2005), 49–103.

22. D. M. Strong, trans., *The Udāna, or the Solemn Utterances of the Buddha* (Oxford: Pali Text Society, 1994), I:10.

23. *Mālunkyāputta Sutta, Sutta Nipāta*, in *The Middle-Length Discourse of the Buddha*, trans. Bhikkhu Ñāṇamoli and Bhikkhu Bodhi (Boston: Wisdom, 1995), XXXV.95.

24. *Saṃyutta Nikāya*, V. 143, in *The Connected Discourses of the Buddha*, trans. Bhikkhu Bodhi (Boston: Wisdom, 2000), I:135; *Dīgha Nikāya*, I.263; Paravahera Vajirañāṇa, *Buddhist Meditation in Theory and Practice* (Kuala Lumpur, Malaysia: Buddhist Missionary Society, 1975), 364.

25. Tsongkhapa Lobzang Drakpa, *Tsongkhapa's Six Yogas of Naropa*, trans. Glenn H. Mullin (Ithaca, NY: Snow Lion, 1996), 176–80.

26. Stephen LaBerge and Howard Rheingold, *Exploring the World of Lucid Dreaming* (New York: Ballantine, 1990); Stephen LaBerge, "Lucid Dreaming and the Yoga of the Dream State: A Psychophysiological Perspective," in *Buddhism and Science: Breaking New Ground*, ed. B. Alan Wallace (New York: Columbia University Press, 2003), 233–58; Stephen LaBerge, *Lucid Dreaming: A Concise Guide to Awakening in Your Dreams and in Your Life* (Boulder, CO: Sounds True, 2004).

27. Padmasambhava, *Natural Liberation*, 141–61; Tsongkhapa Lobzang Drakpa, *Tsongkhapa's Six Yogas of Naropa*, 172–84; Francisco J. Varela, ed., *Sleeping, Dreaming, and Dying: An Exploration of Consciousness with the Dalai Lama*, trans. Geshe Thupten Jinpa and B. Alan Wallace (Boston: Wisdom, 1997); Wallace, *Genuine Happiness*, 170–95.

28. Tsongkhapa Lobzang Drakpa, *Tsongkhapa's Six Yogas of Naropa*, 173.

29. Padmasambhava, *Natural Liberation*, 114–20; Karma Chagmé, *A Spacious Path to Freedom: Practical Instructions on the Union of Mahāmudrā and Atiyoga*, comm. Gyatrul Rinpoche; trans. B. Alan Wallace (Ithaca, NY: Snow Lion, 1998), 85–101.

30. Jay L. Garfield, trans., *The Fundamental Wisdom of the Middle Way: Nāgārjuna's Mūlamadhyamakakārikā* (New York: Oxford University Press, 1995), VIII; Tsongkha-pa, *The Great Treatise on the Stages of the Path to Enlightenment*, 3:Part 2; Geshe Gedün Lodrö, *Walking Through Walls: A Presentation of Tibetan Meditation*, trans. and ed. Jeffrey Hopkins (Ithaca, NY: Snow Lion, 1992), Part Two; Gen Lamrimpa, *Realizing Emptiness: Madhyamaka Insight Meditation*, trans. B. Alan Wallace (Ithaca, NY: Snow Lion, 2002); Zajonc, ed., *The New Physics and Cosmology*, 154–55.

31. Zajonc, ed., *The New Physics and Cosmology*, 78.

32. Garfield, trans., *The Fundamental Wisdom of the Middle Way*, V.

33. Zajonc, ed., *The New Physics and Cosmology*, 160.

34. Ibid., 209.

35. Ibid., 27.

36. Garfield, trans., *The Fundamental Wisdom of the Middle Way*, XIX.
37. Zajonc, ed., *The New Physics and Cosmology*, 79.
38. Garfield, trans., *The Fundamental Wisdom of the Middle Way*, V.
39. Zajonc, ed., *The New Physics and Cosmology*, 28–29.
40. Garfield, trans., *The Fundamental Wisdom of the Middle Way*, I, XVII.
41. Zajonc, ed., *The New Physics and Cosmology*, 190–91.
42. Garfield, trans., *The Fundamental Wisdom of the Middle Way*, XXVI.
43. B. Alan Wallace and Shauna Shapiro, "Mental Balance and Well-Being: Building Bridges Between Buddhism and Western Psychology." *American Psychologist* 161, no. 7 (Oct. 2006): 690–701.
44. Zajonc, ed., *The New Physics and Cosmology*, 46–49.
45. Karma Chagmé, *Naked Awareness: Practical Teachings on the Union of Mahamudra and Dzogchen*, comm. Gyatrul Rinpoche; trans. B. Alan Wallace (Ithaca, NY: Snow Lion, 2000), 239.
46. A. von Staël-Holstein, *Kāśyapaparivarta, A Mahāyāna Sūtra of the Ratnakūṭa Class Edited in the Original Sanskrit, in Tibetan, and in Chinese* (Tokyo, 1977), Section 43, 102–3; cited in Tsong-kha-pa, *The Great Treatise on the Stages of the Path to Enlightenment*, 3:344.
47. Zajonc, ed., *The New Physics and Cosmology*, 92.
48. Karma Chagmé, *Naked Awareness*, 228; H. H. the Dalai Lama, Dzong-ka-ba, and Jeffrey Hopkins, *Yoga Tantra*, 112.
49. Tsong-kha-pa, *The Great Treatise on the Stages of the Path to Enlightenment*, 3:354–57.
50. Tsongkhapa Lobzang Drakpa, *Tsongkhapa's Six Yogas of Naropa*, 182–84.
51. William James, "Does Consciousness Exist?" in *The Writings of William James*, ed. John J. McDermott (1904; reprint, Chicago: University of Chicago Press, 1977), 177–78; William James, "The Notion of Consciousness," in *The Writings of William James*, 184–94.
52. Michael B. Mensky, "Concept of Consciousness in the Context of Quantum Mechanics," *Physics—Uspekhi* 48, no. 4 (2005): 404–5.
53. Ibid., 400.
54. DK80b1. This is a reference from the Diels-Kranz numbering system, which catalogues quotations from the Presocratics. See Hermann Diels and Walther Kranz, *Die Fragmente der Vorsokratiker* (Zurich: Weidmann, 1985).
55. DK80a19.
56. Daniel Dennett, *Consciousness Explained* (Boston: Little, Brown, 1991), 72.
57. B. Georges Dreyfus, *Recognizing Reality: Dharmakīrti's Philosophy and Its Tibetan Interpretations* (Delhi, India: Sri Satguru Publications, 1997).
58. Tsong-kha-pa, *The Great Treatise on the Stages of the Path to Enlightenment*, 3:178.
59. William Shakespeare, *The Tempest*, London: about 1610; Prospero in Act IV, Scene I, lines 148–158.

## 9. Perfect Symmetry

1. Andrei Linde, "Inflation, Quantum Cosmology and the Anthropic Principle," in *Science and Ultimate Reality: Quantum Theory, Cosmology and Complexity, Hon-*

oring *John Wheeler's 90th Birthday*, ed. John D. Barrow, Paul C.W. Davies, and Charles L. Harper Jr. (Cambridge: Cambridge University Press, 2004), 449; Paul Davies, "That Mysterious Flow," *Scientific American* 16, no. 1 (2006): 6–11.

2. Andrei Linde, "Choose Your Own Universe," in *Spiritual Information: 100 Perspectives on Science and Religion*, ed. Charles L. Harper Jr. (West Conshohocken, PA: Templeton Foundation Press, 2005), 139.

3. Ibid.

4. For an early philosophical argument for the illusory nature of time, see John Ellis McTaggart, "The Unreality of Time," *Mind* 17 (1908): 456–73. For a current scientific perspective on this problem, see Paul Davies, *About Time: Einstein's Unfinished Revolution* (New York: Simon and Schuster, 1995).

5. K. C. Cole, *The Hole in the Universe: How Scientists Peered Over the Edge of Emptiness and Found Everything* (New York: Harcourt, 2001), 250.

6. Ibid., 4.

7. H. H. the Dalai Lama, *Dzogchen: The Heart Essence of the Great Perfection*, trans. Geshe Thupten Jinpa and Richard Barron (Ithaca, NY: Snow Lion, 2000), 48–49.

8. Düdjom Lingpa, *The Vajra Essence: From the Matrix of Pure Appearances and Primordial Consciousness, a Tantra on the Self-Originating Nature of Existence*, trans. B. Alan Wallace (Alameda, CA: Mirror of Wisdom, 2004), 255.

9. Padmasambhava, *Natural Liberation: Padmasambhava's Teachings on the Six Bardos*, comm. Gyatrul Rinpoche; trans. B. Alan Wallace (Boston: Wisdom, 1998), 62.

10. Düdjom Lingpa, *The Vajra Essence*, 251.

11. Ibid., 248.

12. Patrul Rinpoche, *The Words of My Perfect Teacher*, trans. Padmakara Translation Group (Walnut Creek, CA: AltaMira Press, 1998); Khenpo Ngawang Pelzang, *A Guide to the Words of My Perfect Teacher* (Boston: Shambhala, 2004).

13. For an authoritative, concise explanation of the Great Perfection on which this summary is partly based, see H. H. Dudjom Rinpoche, *Extracting the Quintessence of Accomplishment: Oral Instructions for the Practice of Mountain Retreat Expounded Simply and Directly in Their Essential Nakedness* (Corralitos, CA: Vajrayana Foundation, reprint, 1998).

14. For instructions on ascertaining pristine awareness while sleeping, see Padmasambhava, *Natural Liberation*, 161–68, and Tsongkhapa Lobzang Drakpa, *Tsongkhapa's Six Yogas of Naropa*, trans. Glenn H. Mullin (Ithaca, NY: Snow Lion, 1996), 200–8.

15. Düdjom, *The Vajra Essence*, 261.

16. Ibid., 307–8.

17. Ibid., 317–18.

18. *Saṃyutta-nikāya* 2:36, cited in Bhikkhu Ñāṇamoli, *The Life of the Buddha: According to the Pali Canon* (Kandy, Sri Lanka: Buddhist Publication Society, 1992), 206.

19. D. M. Strong, trans., *The Udāna, or the Solemn Utterances of the Buddha* (Oxford: Pali Text Society, 1994), 68–69.

20. Augustine, *The Trinity*, trans. Stephen McKenna (Washington, DC: Catholic University of America Press, 1962), I:20; Augustine, *The Confessions*, trans. Maria Boulding (Hyde Park, NY: New City Press, 1997), 33; John Burnaby, *Amor Dei: A Study of the Religion of St. Augustine* (1938; reprint, Norwich: Canterbury Press, 1991), 48.

21. René Descartes, *Discourse on the Method*, trans. Laurence J. Lafleur (New York: Bobbs-Merrill, 1960), VI:62, 45.

22. Edward O. Wilson, *Consilience: The Unity of Knowledge* (New York: Knopf, 1998); Daniel C. Dennett, *Breaking the Spell: Religion as a Natural Phenomenon* (New York: Viking, 2006).

23. Martin Seligman, *Authentic Happiness: Using the New Positive Psychology to Realize Your Potential for Lasting Fulfillment* (New York: Free Press, 2004).

24. http://www.vision.org/trdl/2000/trdl000123.html.

25. R. Shorto, "Belief by the Numbers," *New York Times Magazine*, January 7, 1997; Sharon Begley, "Science Finds God," *Newsweek*, July 20, 1998.

26. Donald F. Duclow, "Divine Nothingness and Self-Creation in John Scotus Eriugena," *The Journal of Religion* 57, no. 2 (April 1977); H. Lawrence Bond, *On Nicholas of Cusa: Selected Spiritual Writings* (New York: Paulist Press, 1997); Daniel C. Matt, "*Ayin*: The Concept of Nothingness in Jewish Mysticism," in *The Problem of Pure Consciousness: Mysticism and Philosophy*, ed. Robert K.C. Forman (New York: Oxford University Press, 1990); Daniel C. Matt, *The Essential Kabbalah: The Heart of Jewish Mysticism* (San Francisco: HarperSanFrancisco, 1995).

27. Stephen Jay Gould, *Rocks of Ages: Science and Religion in the Fullness of Life* (New York: Ballantine, 1999); Langdon Gilkey, *Creationism on Trial* (Minneapolis: Winston Press, 1985), 108–16.

28. Vasubandhu, *Abhidharmakośabhāṣyam*, French trans. Louis de La Vallée Poussin; English trans. Leo M. Pruden (Berkeley: Asian Humanities Press, 1991), IV:1.

29. John Polkinghorne, *Belief in God in an Age of Science* (New Haven: Yale University Press, 2003); Ian G. Barbour, *When Science Meets Religion* (San Francisco: HarperSanFrancisco, 2000); Nancey Murphy, *Theology in the Age of Scientific Reasoning* (Ithaca, NY: Cornell University Press, 1990).

30. His Holiness the Dalai Lama, *The Universe in a Single Atom: The Convergence of Science and Spirituality* (New York: Morgan Road Books, 2005); B. Alan Wallace, *Contemplative Science: Where Buddhism and Neuroscience Converge* (New York: Columbia University Press, 2006).

# BIBLIOGRAPHY

Adler, Stephen L. "Why Decoherence Has Not Solved the Measurement Problem: A Response to P. W. Anderson." *Studies in History and Philosophy of Science* 34 (2003): 135–42.

Atmanspacher, Harald. "Mind and Matter as Asymptotically Disjoint, Inequivalent Representations with Broken Time-Reversal Symmetry." *BioSystems* 68 (2003): 19–30.

Atmanspacher, Harald and Fred Kronz. "Relative Onticity." In *On Quanta, Mind, and Matter*, ed. H. Atmanspacher, A. Amann, and U. Müller-Herold. Dordrecht: Kluwer, 1999.

Atmanspacher, Harald and Hans Primas. "The Hidden Side of Wolfgang Pauli." *Journal of Consciousness Studies* 3, no. 2 (1996): 112–26.

——. "Pauli's Ideas on Mind and Matter in the Context of Contemporary Science." *Journal of Consciousness Studies* 13, no. 3 (2006): 5–50.

Ayer, Alfred Jules. *Language, Truth and Logic.* 2nd ed. London: Gollancz, 1946.

Bacon, Francis. *Novum Organum.* Trans. and ed. P. Urbach and J. Gibson. Peru, IL: Open Court, 1994.

Barbour, Ian G. *When Science Meets Religion.* San Francisco: HarperSanFrancisco, 2000.

Barrow, John D. "Outward Bound." In *Spiritual Information: 100 Perspectives on Science and Religion*, ed. Charles L. Harper Jr. West Conshohocken, PA: Templeton Foundation Press, 2005.

Barrow, John D. and Frank J. Tipler. *The Anthropic Cosmological Principle.* Oxford: Oxford University Press, 1996.

Begley, Sharon. "Science Finds God." *Newsweek* (July 20, 1998).

Bekenstein, Jacob D. "Information in the Holographic Universe." *Scientific American* 9, no. 2 (August 2003).

Bikkhu Bodhi, trans. "Satipaṭṭhāna Sutta." In *The Middle-Length Discourses of the Buddha.* Boston: Wisdom, 1995.

Bitbol, Michel. "Materialism, Stances, and Open-Mindedness." In *Images of Empiricism: Essays on Science and Stances, with a Reply from Bas van Fraassen*, ed. Bradley Monton. Oxford: Oxford University Press, 2007.

——. *Schrödinger's Philosophy of Quantum Mechanics*. New York: Kluwer, 1995.

Bohm, David. "A New Theory of the Relationship of Mind and Matter." *Philosophical Psychology* 3 (1990): 271–86.

Bohr, Niels. *Atomic Theory and the Description of Nature*. Cambridge: Cambridge University Press, 1934.

——. *Essays 1958–1962 on Atomic Physics and Human Knowledge*. New York: Wiley, 1963.

Bond, H. Lawrence, trans. *Nicholas of Cusa: Selected Spiritual Writings*. New York: Paulist Press, 1997.

Bousso, Raphael. "The Holographic Principle." *Reviews of Modern Physics* 74 (2002): 825–74.

Brukner, Časlav and Anton Zeilinger. "Information and Fundamental Elements of the Structure of Quantum Theory." In *Time, Quantum and Information*, ed. Lutz Castell and Otfried Ischebeck. Berlin: Springer Verlag, 2003, 323–55.

Bruner, Jerome. *Beyond the Information Given: Studies in the Psychology of Knowing*. New York: Norton, 1973.

Buddhaghosa. *The Path of Purification*. Trans. Ñāṇamoli Bhikkhu. Kandy: Buddhist Publication Society, 1979.

Burrow, J. W., ed. *Charles Darwin: The Origin of Species*. London: Penguin, 1968.

Chalmers, David J. *Conscious Mind: In Search of a Fundamental Theory*. New York: Oxford University Press, 1996.

Churchland, Paul. *Matter and Consciousness: A Contemporary Introduction to the Philosophy of Mind*. Rev. ed. Cambridge, MA: MIT Press, 1990.

Cole, K. C. *The Hole in the Universe: How Scientists Peered Over the Edge of Emptiness and Found Everything*. New York: Harcourt, 2001.

Cook, David. *Probability and Schrödinger's Mechanics*. Hackensack, NJ: World Scientific, 2003.

Corbetta, Maurizio and Gordon L. Shulman. "Control of Goal-Directed and Stimulus-Driven Attention in the Brain." *Nature Reviews Neuroscience* 3 (March 2002): 210–15.

Curd, Martin and J. A. Cover. "The Duhem-Quine Thesis and Underdetermination." In *Philosophy of Science*, ed. Martin Curd and J. A. Cover. New York: Norton, 1998.

d'Espagnat, Bernard. "Aiming at Describing Empirical Reality." In *Potentiality, Entanglement, and Passion-at-a-Distance*, ed. R. S. Cohen, M. Horne, and J. Stachel. Dordrecht: Kluwer, 1997.

——. "Concepts of Reality." In *On Quanta, Mind, and Matter*, ed. H. Atmanspacher, A. Amann, and U. Müller-Herold. Dordrecht: Kluwer, 1999, 249–70.

H. H. the Dalai Lama. *Dzogchen: The Heart Essence of the Great Perfection*. Trans. Geshe Thupten Jinpa and Richard Barron. Ithaca, NY: Snow Lion, 2000.

——. *The Universe in a Single Atom: The Convergence of Science and Spirituality*. New York: Morgan Road, 2005.

H. H. the Dalai Lama, Dzong-ka-ba, and Jeffrey Hopkins. *Yoga Tantra: Paths to Magical Feats*. Ithaca, NY: Snow Lion, 2005.

Damasio, Antonio R. *The Feeling of What Happens: Body and Emotion in the Making of Consciousness*. New York: Harcourt, 1999.

——. *Looking for Spinoza: Joy, Sorrow, and the Feeling Brain*. Orlando, FL: Harcourt, 2003.

Davies, Paul C.W. *About Time: Einstein's Unfinished Revolution*. New York: Simon & Schuster, 1995.

——. "An Overview of the Contributions of John Archibald Wheeler." In *Science and Ultimate Reality: Quantum Theory, Cosmology and Complexity, Honoring John Wheeler's 90th Birthday*, ed. John D. Barrow, Paul C.W. Davies, and Charles L. Harper Jr. Cambridge: Cambridge University Press, 2004, 3–26.

——. "Particles Do Not Exist." In *Quantum Theory of Gravity*, ed. S. M. Christensen. New York: Adam Hilger, 1984.

——. "That Mysterious Flow." *Scientific American* 16, no. 1 (2006): 6–11.

Dennett, Daniel C. *Breaking the Spell: Religion as a Natural Phenomenon*. New York: Viking, 2006.

——. *Consciousness Explained*. Boston: Little, Brown, 1991.

Descartes, René. *Discourse on the Method*. Trans. Laurence J. Lafleur. New York: Bobbs-Merrill, 1960.

d'Espagnat, Bernard. *Veiled Reality: An Analysis of Present-Day Quantum Mechanical Concepts*. Reading, MA: Addison-Wesley, 1995.

DeWitt, Bryce S. and Neill Graham, eds. *The Many-Worlds Interpretation of Quantum Mechanics*. Princeton: Princeton University Press, 1973.

Diels, Hermann and Walther Kranz. *Die Fragmente der Vorsokratiker*. Zurich: Weidmann, 1985.

Dreyfus, B. Georges. *Recognizing Reality: Dharmakīrti's Philosophy and Its Tibetan Interpretations*. Delhi, India: Sri Satguru Publications, 1997.

Duclow, Donald F. "Divine Nothingness and Self-Creation in John Scotus Eriugena." *The Journal of Religion* 57, no. 2 (April 1977).

Düdjom Lingpa. *The Vajra Essence: From the Matrix of Pure Appearances and Primordial Consciousness, a Tantra on the Self-Originating Nature of Existence*. Trans. B. Alan Wallace. Alameda, CA: Mirror of Wisdom, 2004.

H. H. Dudjom Rinpoche. *Extracting the Quintessence of Accomplishment: Oral Instructions for the Practice of Mountain Retreat Expounded Simply and Directly in Their Essential Nakedness*. Corralitos, CA: Vajrayana Foundation, 1998.

Dyson, Freeman J. "Thought-Experiments in Honor of John Archibald Wheeler." In *Science and Ultimate Reality: Quantum Theory, Cosmology and Complexity, Honoring John Wheeler's 90th Birthday*, ed. John D. Barrow, Paul C.W. Davies, and Charles L. Harper Jr. Cambridge: Cambridge University Press, 2004.

Easlea, Brian. *Witch-Hunting, Magic and the New Philosophy: An Introduction to Debates of the Scientific Revolution 1450–1750*. Brighton, NJ: Humanities, 1980.

Einstein, Albert. *Ideas and Opinions*. New York: Crown, 1954.

——. *Relativity: The Special and the General Theory*. New York: Three Rivers, 1961.

Ellis, George F. R. "Progress in Scientific and Spiritual Understanding." In *Spiritual Information: 100 Perspectives on Science and Religion*, ed. Charles L. Harper Jr. West Conshohocken, PA: Templeton Foundation Press, 2005.

——. "True Complexity and Its Associated Ontology." In *Science and Ultimate Reality: Quantum Theory, Cosmology and Complexity, Honoring John Wheeler's 90th Birthday*, ed. John D. Barrow, Paul C.W. Davies, and Charles L. Harper Jr. Cambridge: Cambridge University Press, 2004.

Enz, C. P. "On Preparata's Theory of a Super Radiant Phase Transition." *Helvetica Physica Acta* 70 (1997): 141–53.

Everett, Hugh. "Short Article." *Reviews of Modern Physics* 29 (1957): 454.

Feynman, Richard. *The Character of Physical Law*. Cambridge, MA: MIT Press, 1983.

Feynman, Richard, R. B. Leighton, and M. Sands. *The Feynman Lectures on Physics*. Reading, MA: Addison-Wesley, 1963.

Finkelstein, David Ritz. "Emptiness and Relativity." in *Buddhism and Science: Breaking New Ground*, ed. B. Alan Wallace. New York: Columbia University Press, 2003.

——. "Ur Theory and Space-Time Structure." In *Time, Quantum and Information*, ed. Lutz Castell and Otfried Ischebeck. Berlin: Springer Verlag, 2003.

Flanagan, Owen. *The Problem of the Soul: Two Visions of Mind and How to Reconcile Them*. New York: Basic Books, 2002.

Forman, Robert K.C. *Mysticism, Mind, Consciousness*. Albany: State University of New York Press, 1999.

Forman, Robert K.C., ed. *The Innate Capacity: Mysticism, Psychology, and Philosophy*. New York: Oxford University Press, 1998.

——. *The Problem of Pure Consciousness: Mysticism and Philosophy*. New York: Oxford University Press, 1990.

Forsee, Aylesa. *Albert Einstein, Theoretical Physicist*. New York: Macmillan, 1963.

Freud, Sigmund. *New Introductory Lectures on Psychoanalysis*. Trans. and ed. James Strachey. New York: Norton, 1989.

Garfield, Jay L., trans. *The Fundamental Wisdom of the Middle Way: Nāgārjuna's Mūlamadhyamakakārikā*. New York: Oxford University Press, 1995.

Genz, Henning. *Nothingness: The Science of Empty Space*. Trans. Karin Heusch. Cambridge, MA: Perseus, 1999.

Geshe Gedün Lodrö. *Walking Through Walls: A Presentation of Tibetan Meditation*. Trans. and ed. Jeffrey Hopkins. Ithaca, NY: Snow Lion, 1992.

Gethin, R.M.L. *The Buddhist Path to Awakening*. Oxford: Oneworld, 2001.

Ginzburg, Vitaly L. *About Science, Myself, and Others*. Bristol: Institute of Physics Publications, 2005.

Greene, Brian. *The Elegant Universe: Superstrings, Hidden Dimensions, and the Quest for the Ultimate Theory*. New York: Norton, 1999.

Greenstein, George. *The Symbiotic Universe: Life and Mind in the Cosmos*. New York: Morrow, 1988.

Greenstein, George and Arthur G. Zajonc. *The Quantum Challenge: Modern Research on the Foundations of Quantum Mechanics*. Boston: Jones and Bartlett, 1997.

Harrison, E. R. *Cosmology: The Scientific Universe*. New York: Cambridge University Press, 1981.

Heisenberg, Werner. *Daedalus: Journal of the American Academy of Arts and Sciences* 87 (1958): 95.

——. *Physics and Philosophy.* London: Penguin, 1989.

Husserl, Edmund. *Ideas Pertaining to a Pure Phenomenology and to a Phenomenological Philosophy.* Vol. 1. Trans. Ted E. Klein and William E. Pohl. Boston: M. Nijhoff, 1980.

James, William. "Does Consciousness Exist?" In *The Writings of William James,* ed. John J. McDermott. Chicago: University of Chicago Press, 1977.

——. "The Notion of Consciousness." In *The Writings of William James,* ed. J. J. McDermott. 1905; reprint, Chicago: University of Chicago Press, 1977.

——. *The Principles of Psychology.* 1890; reprint, New York: Dover, 1950.

——. *The Varieties of Religious Experience: A Study in Human Nature.* New York: Longmans, Green, 1902.

——. "A World of Pure Experience." In *The Writings of William James,* ed. John J. McDermott. 1912; reprint, Chicago: University of Chicago Press, 1977, 194–214.

Jamgön Kongtrul Lodrö Tayé. *Buddhist Ethics.* Trans. The International Translation Committee. Ithaca, NY: Snow Lion, 1998.

——. *Myriad Worlds: Buddhist Cosmology in Abhidharma, Kālacakra and Dzog-chen.* Trans. and ed. The International Translation Committee. Ithaca, NY: Snow Lion, 1995.

Jibu, Mari and Kunio Yasue. *Quantum Brain Dynamics and Consciousness—An Introduction.* Amsterdam: John Benjamins, 1995.

Jung, Carl G. *The Collected Works of C. G. Jung.* Vol. 14. Mysterium Conjunctionis. 2nd ed. Princeton: Princeton University Press, 1970.

Kamalaśīla. *First Bhavanakrama.* Ed. G. Tucci. In *Minor Buddhist Texts, Part II.* Rome: Istituto italiano per il Medio ed Estremo Oriente, 1958.

Kant, Immanuel. *Metaphysical Foundations of Natural Science.* Trans. James Ellington. 1786; reprint, Indianapolis: Bobbs-Merrill, 1970.

Karma Chagmé. *Naked Awareness: Practical Teachings on the Union of Mahamudra and Dzogchen.* Commentary by Gyatrul Rinpoche. Trans. B. Alan Wallace. Ithaca, NY: Snow Lion, 2000.

——. *A Spacious Path to Freedom: Practical Instructions on the Union of Mahāmudrā and Atiyoga.* Commentary by Gyatrul Rinpoche. Trans. B. Alan Wallace. Ithaca, NY: Snow Lion, 1998.

Khenpo Ngawang Pelzang. *A Guide to the Words of My Perfect Teacher.* Boston: Shambhala, 2004.

Klaus, Michael Meyer-Abich. "Science and Its Relation to Nature in C. F. von Weizsäcker's 'Natural Philosophy.'" In *Time, Quantum and Information,* ed. Lutz Castell and Otfried Ischebeck. Berlin: Springer Verlag, 2003.

Klein, Anne C. "Mental Concentration and the Unconditioned: A Buddhist Case for Unmediated Experience." In *Paths to Liberation: The Mārga and Its Transformations in Buddhist Thought,* ed. Robert E. Buswell Jr. and Robert M. Gimello. Delhi: Motilal Banarsidass, 1992, 269–308.

Koch, Christof. *The Quest for Consciousness: A Neurobiological Approach.* Englewood, CO: Roberts, 2004.

Koestler, Arthur. *The Ghost in the Machine*. New York: Macmillan, 1967.

LaBerge, Stephen. *Lucid Dreaming: A Concise Guide to Awakening in Your Dreams and in Your Life*. Boulder, CO: Sounds True, 2004.

——. "Lucid Dreaming and the Yoga of the Dream State: A Psychophysiological Perspective." In *Buddhism and Science: Breaking New Ground*, ed. B. Alan Wallace. New York: Columbia University Press, 2003, 233–58.

LaBerge, Stephen and Howard Rheingold. *Exploring the World of Lucid Dreaming*. New York: Ballantine, 1990.

Lamrimpa, Gen. *Realizing Emptiness: Madhyamaka Insight Meditation*. Trans. B. Alan Wallace. Ithaca, NY: Snow Lion, 2002.

Langer, Susanne. *Philosophy in a New Key: A Study in the Symbolism of Reason, Rite, and Art*. 3rd ed. Cambridge, MA: Harvard University Press, 1978.

Laughlin, Robert B. *A Different Universe: Reinventing Physics from the Bottom Down*. New York: Basic Books, 2005.

Linde, Andre. "Choose Your Own Universe." In *Spiritual Information: 100 Perspectives on Science and Religion*, ed. Charles L. Harper Jr. West Conshohocken, PA: Templeton Foundation Press, 2005.

——. "Inflation, Quantum Cosmology and the Anthropic Principle." In *Science and Ultimate Reality: Quantum Theory, Cosmology and Complexity, Honoring John Wheeler's 90th Birthday*, ed. John D. Barrow, Paul C.W. Davies, and Charles L. Harper Jr. Cambridge: Cambridge University Press, 2004, 426–58.

Lockwood, Michael. *Mind, Brain and the Quantum*. Oxford: Basil Blackwell, 1989.

Lyons, William. *The Disappearance of Introspection*. Cambridge, MA: MIT Press, 1986.

Machamer, Peter, ed. *The Cambridge Companion to Galileo*. Cambridge: Cambridge University Press, 1998.

Matt, Daniel C. "*Ayin*: The Concept of Nothingness in Jewish Mysticism." In *The Problem of Pure Consciousness: Mysticism and Philosophy*, ed. Robert K.C. Forman. New York: Oxford University Press, 1990.

——. *The Essential Kabbalah: The Heart of Jewish Mysticism*. San Francisco: HarperSanFrancisco, 1995.

McMullin, Ernan. "Enlarging the Known World." In *Physics and Our View of the World*, ed. Jan Hilgevoord. Cambridge: Cambridge University Press, 1994, 79–113.

McTaggart, John Ellis. "The Unreality of Time." *Mind* 17 (1908): 456–73.

Mensky, Michael B. "Concept of Consciousness in the Context of Quantum Mechanics." *Physics—Uspekhi* 48, no. 4 (2005): 389–409.

——. "Quantum Mechanics: New Experiments, New Applications, and New Formulations of Old Questions." *Physics—Uspekhi* 43, no. 6 (2000): 585–600.

——. *Quantum Measurements and Decoherence: Models and Phenomenology*. Dordrecht: Kluwer, 2000.

Mermin, N. David. "What Is Quantum Mechanics Trying to Tell Us?" *American Journal of Physics* 66 (1998): 753–67.

Metzinger, Thomas, ed. *Neural Correlates of Consciousness: Empirical and Conceptual Questions*. Cambridge, MA: MIT Press, 2000.

Murphy, Nancey. *Theology in the Age of Scientific Reasoning*. Ithaca, NY: Cornell University Press, 1990.

Nārada Mahā Thera. *A Manual of Abhidhamma*. 4th ed. Kuala Lumpur, Malaysia: Buddhist Missionary Society, 1979.

Newton, Isaac. *The Principia: Mathematical Principles of Natural Philosophy, I*. Trans. Bernard Cohen and Anne Whitman. Berkeley: University of California Press, 1999.

Pa-Auk Tawya Sayadaw. *Knowing and Seeing*. Kuala Lumpur, Malaysia: WAVE Publications, 2003.

Padmasambhava. *Natural Liberation: Padmasambhava's Teachings on the Six Bardos*. Commentary by Gyatrul Rinpoche. Trans. B. Alan Wallace. Boston: Wisdom, 1998.

Paravahera Vajirañāṇa. *Buddhist Meditation in Theory and Practice*. Kuala Lumpur, Malaysia: Buddhist Missionary Society, 1975.

Patrul Rinpoche. *The Words of My Perfect Teacher*. Trans. Padmakara Translation Group. Walnut Creek, CA: AltaMira Press, 1998.

Penrose, Roger. *The Emperor's New Mind*. Oxford: Oxford University Press, 1989.

——. *Shadows of the Mind*. New York: Vintage, 1995.

Polkinghorne, John. *Belief in God in an Age of Science*. New Haven: Yale University Press, 2003.

Postman, L., J. Bruner, and R. Walk. "The Perception of Error." *British Journal of Psychology* 42 (1951): 1–10.

Putnam, Hilary. *The Many Faces of Realism*. La Salle, IL: Open Court, 1987.

——. *Realism with a Human Face*. Ed. James Conant. Cambridge, MA: Harvard University Press, 1990.

Quine, Willard Van Orman. *From Stimulus to Science*. Cambridge, MA: Harvard University Press, 1995.

——. "On Empirically Equivalent Systems of the World." *Erkenntnis* 9 (1975): 313–28.

——. *Ontological Relativity and Other Essays*. New York: Columbia University Press, 1969.

Remy, Lorraine Nicolas. *Demonolatry*. Trans. E. A. Ashwin. London: University Books, 1930.

Ryle, Gilbert. *The Concept of Mind*. London: Hutchinson, 1963.

Riordan, Michael. "Science Fashions and Scientific Fact." *Physics Today* 56, no. 8 (August 2003): 50.

Saddhatissa, Hammalawa. *Buddhist Ethics*. Boston: Wisdom, 1997.

Saṃyutta Nikāya, V:143. In *The Connected Discourses of the Buddha*, trans. Bhikkhu Bodhi. Boston: Wisdom, 2000.

Schlipp, Paul Arthur. *Albert Einstein: Philosopher-Scientist*. Evanston, IL: Library of Living Philosophers, 1949.

Schrödinger, Erwin. *The Interpretation of Quantum Mechanics*. Woodbridge, CN: Ox Bow Press, 1995.

——. *Nature and the Greeks*. New York: Columbia University Press, 1954.

Searle, John R. *Consciousness and Language*. Cambridge: Cambridge University Press, 2002.

——. *Mind: A Brief Introduction*. New York: Oxford University Press, 2004.

——. *The Rediscovery of the Mind*. Cambridge, MA: MIT Press, 1994.

Shakespeare, William. *The Tempest*. London: about 1610.

Shear, Jonathan. *The Inner Dimension: Philosophy and the Experience of Consciousness*. New York: Peter Lang, 1990.

Shoemaker, Sydney. *The First-Person Perspective and Other Essays*. Cambridge: Cambridge University Press, 1996.

Shorto, Russell. "Belief by the Numbers." *New York Times Magazine*, January 7, 1997, 60–61.

Skinner, B. F. *Science and Human Behavior*. New York: Macmillan, 1953.

Sprat, Thomas. *The History of the Royal Society of London*. ed. J. I. Cape and H. W. Jones. London: Routledge, 1959.

Stich, Stephen. *From Folk Psychology to Cognitive Science: The Case Against Belief*. Cambridge, MA: Bradford, 1983.

Strong, D. M., trans. *The Udāna, or the Solemn Utterances of the Buddha*. Oxford: Pali Text Society, 1994.

Susskind, Leonard. "Black Holes and the Information Paradox." *Scientific American* 276, no. 4 (April 1997): 52–57.

———. "The World as a Hologram." *Journal of Mathematical Physics* 36 (1995): 6377–6396.

Taylor, Charles. *Sources of the Self: The Making of the Modern Identity*. Cambridge, MA: Harvard University Press, 1989.

Taylor, Edwin and John A. Wheeler. *Space-Time Physics*. 2nd ed. New York: W. H. Freeman, 1992.

Tsongkhapa. *The Great Treatise on the Stages of the Path to Enlightenment*. Vol. 3. Ithaca, NY: Snow Lion, 2002.

———. *Tsongkhapa's Six Yogas of Naropa*. Trans. Glenn H. Mullin. Ithaca, NY: Snow Lion, 1996.

Van Fraassen, Bas. *The Empirical Stance*. New Haven: Yale University Press, 2002.

———. "From Vicious Circle to Infinite Regress and Back Again." In *Proceedings of the 1992 Biennial Meeting of the Philosophy of Science Association*, vol. 2, ed. D. Hull, M. Forbes, and K. Okruhlick. East Lansing, Michigan, 1993.

———. *The Scientific Image*. New York: Oxford University Press, 1980.

Varela, Francisco J., ed. *Sleeping, Dreaming, and Dying: An Exploration of Consciousness with the Dalai Lama*. Trans. Geshe Thupten Jinpa and B. Alan Wallace. Boston: Wisdom, 1997.

Vasubandhu. *Abhidharmakośabhāṣyam*. French trans. Louis de La Vallée Poussin. English trans. Leo M. Pruden. Berkeley: Asian Humanities Press, 1991.

Ven. Anālayo. "Mindfulness in the Pāli Nikāyas." In *Buddhist Thought and Applied Psychology: Transcending the Boundaries*, ed. D. K. Nauriyal. London: Routledge-Curzon, 2006, 229–49.

von Meyenn, Karl, ed. *Wolfgang Pauli. Wissenschaftlicher Briefwechsel, Band III: 1940–1949*. Berlin, Springer, 1993.

von Staël-Holstein, Alexander. *Kāśyapaparivarta, A Mahāyāna Sūtra of the Ratnakūṭa Class Edited in the Original Sanskrit, in Tibetan, and in Chinese*. Tokyo, 1977.

von Weizsäcker, Carl Friedrich. *The History of Nature*. London: Routledge & Kegan Paul, 1951.

——. *The Unity of Nature*. Trans. Francis J. Zucker. New York Farrar, Straus & Giroux, 1980.

Wallace, B. Alan. *The Attention Revolution: Unlocking the Power of the Focused Mind*. Boston: Wisdom, 2006.

——. *Balancing the Mind: A Tibetan Buddhist Approach to Refining Attention*. Ithaca, NY: Snow Lion, 2005.

——. *Choosing Reality: A Buddhist View of Physics and the Mind*. Ithaca, NY: Snow Lion, 1996.

——. *Contemplative Science: Where Buddhism and Neuroscience Converge*. New York: Columbia University Press, 2007.

——. *Genuine Happiness: Meditation as the Path to Fulfillment*. Hoboken, NJ: Wiley, 2005.

——. *The Taboo of Subjectivity: Toward a New Science of Consciousness*. New York: Oxford University Press, 2000.

——. "Vacuum States of Consciousness: A Tibetan Buddhist View." In *Buddhist Thought and Applied Psychology: Transcending the Boundaries*, ed. D. K. Nauriyal. London: Routledge-Curzon, 2006.

Wallace, B. Alan and Shauna Shapiro. "Mental Balance and Well-Being: Building Bridges Between Buddhism and Western Psychology." *American Psychologist* 161, no. 7 (Oct. 2006): 690–701.

Waterfield, Robin, ed. *The First Philosophers: The Presocratics and Sophists Translated with Commentary*. Oxford: Oxford University Press, 2000.

Watson, John B. *Behaviorism*. 1913; reprint, New York: Norton, 1970.

——. "Psychology as the Behaviorist Views It." *Psychological Review* XX (1913): 158–77.

Wigner, Eugene P. "Physics and the Explanation of Life." *Foundations of Physics* 1 (1970): 35–45.

——. "Remarks on the Mind-Body Question." In *Quantum Theory and Measurement*, ed. John Archibald Wheeler and Wojciech Hubert Zurek. Princeton: Princeton University Press, 1983.

——. *Symmetries and Reflections*. Bloomington: Indiana University Press, 1967.

Wild, John, ed. *Spinoza: Selections*. New York: Charles Scribner's Sons, 1930.

Wilson, Edward O. *Consilience: The Unity of Knowledge*. New York: Knopf, 1998.

Wittgenstein, Ludwig. *On Certainty*. Trans. Denis Paul and G.E.M. Anscombe. San Francisco: Arion Press, 1991.

——. *Philosophical Investigations*. Trans. G.E.M. Anscombe. Oxford: Blackwell, 1958.

Zajonc, Arthur, ed. *The New Physics and Cosmology: Dialogues with the Dalai Lama*. New York: Oxford University Press, 2004.

Zeh, Heinz Dieter. "There Are No Quantum Jumps, Nor Are There Particles." *Physics Letters* A172 (1993): 189–92.

Zeilinger, Anton. "Why the Quantum? 'It' from 'Bit'? A Participatory Universe? Three Far-Reaching Challenges from John Archibald Wheeler and Their Relation to Experiment." In *Science and Ultimate Reality: Quantum Theory, Cosmology and Complexity, Honoring John Wheeler's 90th Birthday*, ed. John D. Barrow, Paul C.W. Davies, and Charles L. Harper Jr. Cambridge: Cambridge University Press, 2004, 201–20.

# INDEX

Buddhism (*continued*)
   mentarity, 117–18; contrast to Christianity, 120; cosmogony, 87, 93–94, 110–14; dependent origination concept, 96, 99; dreaming and waking states, 90–93; hypotheses derived from Tibetan Buddhism, 43–49; lack of independent existence of things, 90, 92–94, 98, 105; and Mensky's speculations, 102–3; parallels in quantum physics/quantum cosmology, 85–88, 95, 110; and paranormal abilities, 99–103, 116–17; quantum concepts, 86–87, 98; validity of introspective observations, 105–6. *See also* karma; meditative practices; *specific types*

Cartesian dualism, 24, 105, 108, 120
Cartesian science, 28–30, 41
*causa sui*, 54
causality: and brain functions, viii, 22, 29, 30; and Cartesian science, 29–30; classes of phenomena with causal effects, 24, 34; and classical physics, 30; Ellis's model of reality, 55–56; and modern physics, 23, 30, 96; natural world equated to the world of physical causality, ix, 23–26; and neural correlates of consciousness, 22; and time, 96–97
Christianity, 13, 118–19
chronon, 87
classical physics. *See* physics, classical
cognitive sciences, viii–ix; assumption of equivalence of mental phenomena and neurophysiological processes, 5, 7; assumption of mechanical explanations for all causal relationships, 29–30; assumptions about dualism and monism, 24; assumptions reinforced by method of inquiry, 5, 8; emergence of the mind sciences, 4–6; "hard problem" of neurophysiology, ix, 23, 75, 105; and ideological hierarchies of knowledge, 11–12, 15; introspection as appropriate tool, 5, 20–

21, 39; marginalization of mental phenomena, 4–5, 105; and moral relativism, 104; problems with objectivist orientation, 104–5; reliance on future discoveries for validating present beliefs, 25; revolution in, 12–15. *See also* consciousness; science of consciousness
colors, existence of, 72
complementarity, 28, 117–21
complementary virtues, 67–68
computers, 51, 74
conditioned existence, world of, 93–94
consciousness: alternative definitions, 5–6; "bracketing" consciousness from its object, 39; and Buddhist cosmogony, 93–94, 110–13; existence independent of matter, 31, 33, 34; "hard problem" of neurophysiology, ix, 23, 75, 105; hypotheses about role in the universe, viii, 31–32, 109; hypotheses derived from Tibetan Buddhism, 43–49; and individual choice, 82–83; influence on brain, 24, 33; and information, 34, 73–76; insufficiency of reductionist hierarchy of knowledge, 56; and many-worlds hypothesis, 81–82, 102–3; need for openness to metaphysical theories, 30–31; and neo-Darwinism, 21; primordial consciousness, 110–14; problems with dualist and materialist theories, 8, 24; and reciprocal influences among fields of knowledge, 14; spiritual awakening, 101, 110, 113, 116–17; "superfluid" state of, 102; and zero-point field, 34. *See also* empirical methods for observing the space of the mind; measurement problem in quantum mechanics; mental phenomena; neural correlates of consciousness; observer; quantum consciousness; science of consciousness; substrate consciousness
conservation of mass-energy, 33–34
constants of nature, 53

Copenhagen school of quantum mechanics, 77, 81
Copernicus, Nicolaus, vii, 2–3, 12
correspondence theory of appearances and physical reality, 51–52
cosmology: anthropic principle, 79–80, 82; Buddhist cosmogony, 87, 93–94, 110–14; and delayed-choice experiment, 78–80; empirical origins of theories, 36–37; and frozen time problem, iii, 108–9; holographic universe, 54–57, 73; need for letting go of assumptions about role of consciousness in the universe, viii, 30–31; participatory universe, 78–79, 88, 109; and time evolution of the universe, 108–9. *See also* quantum cosmology
creative energy, 46
Crick, Francis, 30

Dalai Lama, 85, 95, 99, 110–11
Darwin, Charles, vii, 4, 12, 30
Davies, Paul C. W., 34
death, 116–17
decoherence, 19, 33; defined, 28; and measurement problem, 32–33; and quantum cosmology, 78
deism, 26
delayed-choice experiment, 78–80
Democritus, 52–53
Descartes, René, 28, 105, 118
d'Espagnat, Bernard, 54
DeWitt, Bryce, 78, 81, 108
*dharmadhātu*, 110. *See also* absolute space of phenomena
dimensions: open questions in particle physics, 59. *See also* hidden dimensions
DNA, 30
dream yoga, 91–93, 99–100
dreaming: and Buddhism, 90–93, 115; lucid dreaming, 39, 91–92, 99–100, 116; and neural correlates of consciousness, 22; and observer participancy, 39
dreamless sleep, 46, 47, 92, 102

dualism, 24, 105, 108, 112, 120. *See also* objective reality
Düdjom Lingpa, 111, 112
Dyson, Freeman, 28
Dzogchen tradition of Tibetan Buddhism. *See* Great Perfection (Dzogchen) tradition of Tibetan Buddhism

earth element, 61–62
Einstein, Albert, 40, 72, 74–75, 96
Ellis, George, 55–56
embryo, 48
emergent behavior, 5, 14, 54, 55, 87
emotions: discrepancies between appearances and reality, 44–45
empirical methods for observing the space of the mind, 36–49; and archetypal forms, 61–69; and Baconian vs. Cartesian science, 41; developing a telescope for the mind, 41–43; discrepancies between appearances and reality, 44–45; experiments in quantum consciousness, 85–107; high-energy experiments in consciousness, 58–69; hypotheses derived from Tibetan Buddhism, 43–49; need for training, 39–40, 41, 42–43, 59; parallels in astronomy, 36–37, 47; parallels in mathematics, 39–40; parallels in particle physics, 58–60; questions guiding observations, 43; repeatability, 42; and scientific evaluation of theories, 66–69; semiprivate language of contemplatives, 42, 44; testable hypotheses, 39; traditional resistance to introspection, 37–41; validity of introspective observations, 40–41. *See also* meditative practices
energy: and naturalism, 17; parallels between Buddhism and quantum physics, 86–87; primordial energy, 110. *See also* mass-energy
entanglement, 19, 82, 97
entropy, and information, 73
Enz, Charles, 33

ethics, 60–61, 63, 68, 89, 98, 120
Euclid, 2
Everett, Hugh, 81–84, 102
evolution: and Baconian vs. Cartesian science, 30; human evolution and need for ethics in science, 68–69; neo-Darwinism, 21–23
existence, criteria for, 106. *See also* Buddhism: cosmogony; reality
expectations and perceptions, 40, 52, 71
Extended Everett's Concept, 103
"Extinction into Reality-Itself," 116

Feynman, Richard, 49
Finkelstein, David, 96, 97
fire element, 64
form realm, 65, 94, 101, 110. *See also* archetypes
formless realm, 66, 93–94, 110
Four Applications of Mindfulness, 89–90
"fourth time," 111, 114
Freud, Sigmund, 40, 44
frozen time problem, viii, 108–9
frozen vacuum state, 109

Galileo Galilei, 3, 10, 36–37, 39–40, 118
Gamow, George, 37
general theory of ontological relativity, 70–84; and Buddhism, 98; existence of phenomena in relation to cognitive frame of reference, 70–72, 98, 105; information and consciousness, 73–76; and many-worlds hypothesis, 80–94; philosophical precedents, 70–72; and quantum physics/quantum cosmology, 76–80
general theory of relativity, 23, 30
Gestalt-switch, 41
Ginzburg, Vitaly L., 32
Gnostic tradition, 118
God, 26, 28, 118–20, 121
gravitational waves, 96
gravity, 30. *See also* general theory of relativity

Great Perfection (Dzogchen) tradition of Tibetan Buddhism, 48; cosmogony, 110–14; training in meditative practices, 113–16

happiness, genuine, 63, 64, 69, 98–99, 106, 118–19, 120
"hard problem" of neurophysiology, ix, 23, 75, 105
Hawking, Stephen, 27–28, 28, 73
hedonic pleasure, 99, 119
Heisenberg, Werner, 53
Heisenberg Uncertainty Principle, 34
hidden dimensions: and objective reality, 18–19; of the psyche, 47–48
high-energy experiments in consciousness, 58–60; and scientific evaluation of theories, 66–69; training for, 60–61
history of science, 1–15; competing perspectives, 2; early history, 1–4; emergence of the mind sciences, 4–6; ideas of matter, 52–53; ideological hierarchies of knowledge, 8–12; and idolization of laws of nature, 97; individuals as authorities on the nature of reality, 2; long tradition of philosophical speculation in empirical sciences, 59; mental phenomena as blind spot of traditional science, ix, 6–8, 25; scientific revolutions, vii, 3–4; and theology, 26, 118–20; underdetermination problem, 77
holographic universe, 54–57, 73
Hubble, Edwin Powell, 36
Hubble Space Telescope, 37
Hubble Ultra Deep Field, 37, 47
Humason, Milton, 36–37
Husserl, Edmund, 39
Hut, Piet, 95

ideological hierarchies of knowledge, 8–12
imagination, 50–51, 101
inflation in the early universe, 31

information, 73–76; and black holes, 73; and classes of phenomena with causal effects, 34; and consciousness, 34, 73–76; and general theory of ontological relativity, 73–76; and holographic universe, 55; location of, 51–52; and qualia, 51–52; Wheeler's speculations, 73–74

insight, contemplative, 88–101, 114

introspection: as appropriate method of inquiry for the mind sciences, 5, 20–21, 39; and perception as function of expectation, 40; traditional resistance to, 37–41; validity of introspective observations, 40–41; and zero-point field, 34. *See also* empirical methods for observing the space of the mind; meditative practices

"it from bit" dictum, 74, 75

James, William, 10, 39, 40, 101

Jibu, Mari, 33

*jñāna*, 110. *See also* primordial consciousness

*jñā-prāṇa*, 110. *See also* primordial energy

joy, 44–45, 64

Jung, Carl, 53–54

Kant, Emmanuel, 38

karma, 93, 97, 101, 110, 115

Kepler, Johannes, 3

knowledge, ideological hierarchies, 8–12

Koch, Christof, 25

*kṣaṇa*, 87

language, 55

Large Hadron Collider, 58

life, origins of, 11–12, 32, 56

lifestyle, and training in meditative practices, 60–61, 114, 115

light: delayed-choice experiment, 78–80; lack of mechanical explanation for propagation of, 30; and measurement problem of quantum mechanics, 39, 78–80; quantization of, 86; reciprocal influences of light and matter, 24; and special relativity, 72; and zero-point field, 34

Linde, Andrei, 30–31, 109

Lipperhey, Hans, 36

location of visual images, 50

*loka*, 87

lucid dreaming, 39, 91–92, 99–100, 116

lucid dreamless sleep, 46, 92, 102

Mahāyāna Buddhism, 66, 89, 94, 99

many-worlds interpretation of quantum physics, 81–84, 102–3

mass-energy: and broken symmetries, 109; and classes of phenomena with causal effects, 24; open questions in particle physics, 58–59; parallels between Buddhism and quantum physics, 86–87, 95

mathematics: and classes of phenomena with causal effects, 24; independent existence of, 56–57; insufficiency of reductionist hierarchy of knowledge, 11, 56; and matter, 53, 56; need for professional training in, 39–40; reciprocal influences among fields of knowledge, 14

matter: historical ideas of, 52–53; and Jungian archetypes, 53–54; and mathematics, 53, 56; open questions in particle physics, 58–59; and physical laws, 97; problems with classical concept of, 16–17

*māyā*, 93

McMullin, Ernan, 67

measurement problem in quantum mechanics, vii–viii; and consciousness, 32–33, 75; Copenhagen interpretation, 77, 81; and decoherence, 19, 32–33; defined/described, 38–39; delayed-choice experiment, 78–80; entanglement, 19, 82, 97; and "hard problem" of neurophysiology, ix, 105; many-worlds hypothesis, 81–84; and objective reality, 19;

measurement problem (*continued*)
and philosophical resistance to introspection, 38; Schrödinger on, 86; and selection mechanism, 32

meditative practices: and attention, 89–90; dream yoga, 91–93, 99–100; Four Applications of Mindfulness, 89–90; Great Perfection, 113–16; inquiry into archetypal forms, 61–66; insights into the nature of the mind, 94; insights into the self as agent, 94–95; meditative quiescence, 88–89; and mental imbalances, 64–65, 89, 100; need for training, 42–43, 60–61; and paranormal abilities, 99–103, 116–17; and quantum world of experience, 88–90; repeatability issues, 67; results of, 63, 88–89; and scientific evaluation of theories, 66–69; semi-private language of contemplatives, 42, 44; settling the mind in its natural state, 45–46; unification of quiescence and insight, 98–103; validity of introspective observations, 105–6. *See also* empirical methods for observing the space of the mind

melted vacuum state, 109–10

memory, 33, 38, 48, 71

Mensky, Michael, 31–33, 81–84, 102

mental phenomena, viii, 29–30; acceptable epistemic methods for observing mental phenomena, 20–21; assumption of equivalence of mental phenomena and neurophysiological processes, 5, 7; assumption of mechanical explanations for all causal relationships, 29–30; Baconian vs. Cartesian approach to study of, 28–30; as blind spot in traditional science, ix, 6–8, 25; discrepancies between appearances and reality, 44–45; illusory nature of perceptions, 50–52; imagination, 50–51; and Jungian archetypes, 53–54; lack of existence independent of cognitive frame of reference, 94–95; mental imbalances

alleviated through meditative practices, 64–65, 89, 100; mental vocabulary, 39; and real-time observations, 38; study of mental phenomena marginalized, 4–5, 25. *See also* empirical methods for observing the space of the mind; introspection; neural correlates of consciousness

metaphysics, and the history of science, 59

Michelson-Morley experiment, 17, 23

Middle Way, 94–98, 105–7

mind-body problem, viii–ix, 24, 30, 52, 105. *See also* brain; cognitive sciences; consciousness; neural correlates of consciousness

miracles, 83, 103

modern physics. *See* physics, modern

moral relativism, 104

M-theory, 18–19

natural laws, 54, 93, 97, 109

naturalism, 16–26; natural theory of human consciousness (*see* science of consciousness); natural world equated to objective world, 17–18; natural world equated to the physical world, 16–17; natural world equated to the world of neo-Darwinism, 21–23; natural world equated to the world of physical causality, 23–26; natural world equated to the world of physics, 20–21; problems with concepts of matter and energy, 16–17; and quantum mechanics, 16, 19

neo-Darwinism, 5, 21–23

neural correlates of consciousness, 21–23; assumption of equivalence of mental phenomena and neurophysiological processes, 5, 7; and causality, 24; and dreaming, 22; "hard problem" of neurophysiology, ix, 23, 75, 105; location of visual images, 50; necessary vs. sufficient causes, 22, 25; omission of first-person perspective, 23; unfalsifiability of theory, 23

qualia, 50–52
quantum brain dynamics, 33
quantum consciousness, 85–107;
dreaming and waking states, 90–93;
evaluation of experiences and insights, 103–7; meditative practices,
88–90; parallels between Buddhism
and quantum physics, 85–88; and
paranormal abilities, 99–103; and
quantum relativity, 93–98; unification
of quiescence and insight, 98–103
quantum cosmology, 28, 31; and delayed-choice experiment, 78–80; and entanglement, 97; frozen time problem,
viii, 108–9; and participatory universe, 109
quantum field theory, 24, 33
quantum physics, 7, 67–69; Copenhagen interpretation, 77, 81; delayed-choice experiment, 78–80; entanglement, 19, 82, 97; exclusivist view of,
27; inclusivist view of, 27–28; and information, 75; many-worlds interpretation, 81–84, 102–3; and matter, 53;
parallels between Buddhism and
quantum physics, 85–88, 110; and
problems with different types of naturalism, 16, 19; quantum interference
effects, 29; relation to classical physics, 28, 77–79, 82, 107 (See also decoherence); and role of consciousness
in the universe, viii, 31–32 (See also
quantum cosmology). See also measurement problem in quantum
mechanics
quiescence, meditative, 88–89; and
dream states, 91; unification of quiescence and insight, 98–103
Quine, Willard, 70

reality: competing perspectives in the
history of science, 2, 52–53; and
different types of naturalism, 16–26;
Ellis's model, 55–56; existence of
hidden dimensions, 18–19; existence
of phenomena only in relation to

cognitive frame of reference, 72, 92–94; and frozen time problem, 108;
and holographic universe, 55–56;
illusory nature of perceptions, 50–52;
and primordial consciousness, 112;
and special relativity, 19; and
supernatural phenomena, 18. See also
archetypes; Buddhism: cosmogony;
objective reality
reductionism, 10–11, 32
reincarnation, 48
religion: and background theories, 107;
and history of science, 118–20; and
ideological hierarchies of knowledge,
10–11; reciprocal influences among
fields of knowledge, 14; religious fundamentalism, 8; and supernatural
phenomena, 18. See also Buddhism;
Christianity
Rinpoche, Penor, 117
Riordan, Michael, 67
Roman Catholic Church, 13, 118
rūpa-dhātu, 65. See also form realm
Rutherford, Ernest, 58
Ryle, Gilbert, 41, 44

śamatha, 88
samyag-dṛṣṭi, 89
samyak-saṃkalpa, 89
samyak-smṛti, 89
Sanskrit, 93
Santa Barbara Institute for Consciousness Studies, 43
Schrödinger, Erwin, 16, 25, 86, 93
Schrödinger wave function of the universe, 108
science, traditional: alternative evolution
of sciences, 13–14; background/foreground theories, 76–77, 107; Baconian vs. Cartesian science, 28–29; different types of naturalism, 16–26;
emergence of the mind sciences, 4–6; evaluation of theories, 106, 118–19;
history of (See history of science); ideological hierarchies of knowledge, 8–12; and implications of Mensky's ex-

tension of Everett's many-worlds hypothesis, 83; insufficiency of reductionist hierarchy of knowledge, 11–12, 56; lack of ethics in, 68, 98; materialistic assumptions reinforced by method of inquiry, 5, 8; mental phenomena as blind spot of, ix, 6–8, 25, 104; mutability of theories and laws, 33–34; natural laws as idols, 97; need for testable hypotheses, 14; neo-Darwinism, 5, 21–23; open-minded vs. closed-minded types of skepticism, 48–49; persistence of maladaptive scientific thought, 13; reciprocal effects among sciences, 13; scientific materialism, 5, 7, 10–11, 13, 29, 41; scientific naturalism, 8; and technology, 68. *See also* cognitive sciences; natural laws; naturalism

science of consciousness, 98–103; Baconian vs. Cartesian science, 28–29; and complementary virtues, 67–69; evaluation of theories, 66–69, 98–99, 103–7; hypotheses concerning general theory of ontological relativity, 70–84; hypotheses concerning special theory of ontological relativity, 50–57; hypotheses derived from Tibetan Buddhism, 43–49; need for letting go of assumptions about role of consciousness in the universe, 30–31; need for openness to metaphysical theories, 30–31; observing the space of the mind, 36–49 (*See also* empirical methods for observing the space of the mind; introspection; meditative practices); outline of possible natural theories of human consciousness, 27–35; reciprocal influences among fields of knowledge, 14; repeatability issues, 67; testable hypotheses, 14, 39. *See also* general theory of ontological relativity; special theory of ontological relativity

self, agency of, 94–95

sensory qualia, 50–52

Shakespeare, William, 107

skepticism, open-minded vs. closed-minded varieties, 48–49

sleep. *See* dreaming; dreamless sleep; lucid dreaming; lucid dreamless sleep

*smayak-smṛti*, 89

solipsism, 105, 106

soul, Christian concept of, 118

space-time: and broken symmetries, 109; and Buddhist cosmogony, 93–94, 110–11; and classes of phenomena with causal effects, 24, 34; location of qualia, 52; quantization of, 87; and special relativity, 19

special relativity, 19, 72, 96

special theory of ontological relativity, 50–57; holographic universe, 54–57; ideas of matter, 52–53; illusory world of perception, 50–52; psychophysical coemergence, 53–54

Spinoza, Benedict de, 54, 74

spiritual awakening, 110, 113, 116–17

subjectivity, marginalization of, 4–6, 25, 37–41, 105

substrate, 46–47, 112; dreaming and waking states, 92–93, 100

substrate consciousness, 48; and archetypal forms, 62–63; characteristics of, 45–46, 48; and lucid, dreamless sleep, 102; and primordial consciousness, 112; and quiescence, 101

supernaturalism, 8

superparticles, 59

superstring theory, 18–19

Susskind, Leonard, 55, 110

symmetry: broken symmetries, 47, 109; and classes of phenomena with causal effects, 24; open questions in particle physics, 59; and substrate, 47; and vacuum states, 109

't Hooft, Gerard, 55, 73

Takahashi, Yoshiyuki, 33

technology, 68, 99, 106, 119

telescope, 36–37

*Theaetetus* (Plato), 104